THE MARVELOUS THEORY OF SPECIAL AND GENERAL RELATIVITY

Umberto Piacquadio

INDEX

Preface

It is without too many scientific pretensions that I am about to introduce the reader to the Theory of General and Narrow Relativity that is the subject of the present paper which, however, cannot be fully grasped without at least the fundamental mathematical and physical skills that are learned at the end of the high school course of study, necessarily with a scientific orientation.

I will try to cover the topics, so that the concepts are mainly understood, through rigorous mathematical demonstrations or with the help of graphs and diagrams, with an innovative approach compared to many other treatises that are often too "popularized" for a proper understanding of the theories or alternatively present a high dose of "college math."

I paid special attention to the thought of Relativity already present in Galileo's observations and expounded on the fundamental works of James Clerk Maxwell, Albert Abraham Michelson, Edward Morley, and Hendrik Antoon Lorentz to arrive at the absolutely innovative idea postulated by the Theory of Restricted Relativity.

Once immersed in the new Theory, in the second chapter, I delved analytically into the concepts on time dilation and distance contraction until the solution of the always fascinating paradoxes.

Subsequently, this paper focuses on mass-energy dualities (E=mc2) and gravitational mass-inertial mass.

There is no shortage of "mental experiments" and the use of the Minkowsky diagram for the purpose of making the effects of the theory better understood.

In the discussion of General Relativity, the topics of the curvature of space-time in four dimensions, Riemann geometry, the use of Tensors, the exposition of Einstein's famous Field Equations, Gravitational

Redshift, Light Deflection, Mercury's Perihelion Precession and Gravitational Waves will be addressed.

I thank all those who have been close to me during the writing of this discussion, and with the hope that I have set up the work in such a way that it will be useful to all those who approach the study of the fascinating Theory of Relativity, I am grateful in advance to those who would like to propose improvements or any suggestions.

Umberto Piacquadio

.... to Marina, Alice, Lorenzo and Luca

CHAPTER 1
RELATIVE MOTIONS

1.1 MODERN PHYSICS

The term Physics comes from the Latin *physica* ("nature") in turn derived from the Greek *physiká* ("natural things"), born from *phýsis*.

Physics is a science that aims to find mathematical relationships to be used as physical laws, suitable for representing natural phenomena. Having found the "physical law," it is thus possible to predict the "causal" evolution of the phenomenon under investigation.

When we speak of "law," we are accustomed to thinking of regulative principles of behavior, catalogued in rules issued by appropriate legislative bodies. The people must act and move in full compliance with the "law."

The term "law" in physics has, however, a misuse. It is the "law" that seeks to describe a natural phenomenon, not the natural phenomenon that obeys a "physical law."

Throwing a stone from the window of the house, through the use of simple physical laws it is possible to know the time it takes for the stone to touch the ground or the speed attained.

Many times, simplifications are made to achieve this goal: the braking action of air (friction), the shape of the object, the effect of wind, changes in temperature and pressure, etc. are neglected.

These approximations might make Physics appear to be an approximate science. In reality, the operation of simplification in the mathematical description of a physical phenomenon, if applied judiciously, does not appreciably alter the hoped-for results.

Neglecting the effect of wind, in calculating the time it takes for a stone thrown, from a window on the third floor of a building, to reach the ground, is acceptable.

What is quite different, however, is to calculate the falling time of a feather thrown from the same window, neglecting the effect of wind.

Ultimately with "classical physics" through appropriate "physical laws" it is possible, within the limit of some approximations, to predict the evolution of a physical system (set of constituent elements of a natural phenomenon) known initial conditions.

Classical physics is based on the concept of "determinism."

The evolution of a natural phenomenon, in a deterministic conception, is based on the principle of causality, that is, founded on the relationship of necessity between cause and effect.

If I kick a ball, which we can call a "cause," I can only have as an "effect" a throw at a proper distance.

Up to here everything is simple.

In the early 20th century, however, comes MODERN PHYSICS, which is not so modern since it is more than 100 years old.

Modern physics is divided into two major branches: Einstein's relativity and Quantum physics.

Einstein's relativity, is aimed at large objects (macrocosm), and although it is classified as Modern Physics, it continues to be a "deterministic" theory.

Quantum physics, on the other hand, studies an infinitesimal reality (microcosm), where nature has a very bizarre behavior: it is composed of disordered infinitesimal particles, which are difficult to represent individually, although in the whole then they create a surprising order.

A water molecule consisting of a multitude of infinitesimal elements (atoms, protons, neutrons, electrons, quarks, ...) joined together to form, for example, "the sea," is shown to us of unprecedented order, beauty and elegance.

Quantum physics is an "indeterministic" type of theory, that is, it studies matter in its behavior in much the same way as the output of a number in a roll of the dice.

Although both theories have achieved an unparalleled status in the modeling of physical phenomena, many physicists are still at work in the search for a unified theory that encompasses the study of the infinitely small (Quantum Physics) and the study of the infinitely large (Theory of Relativity).

Recall that the two fields of modern physics are contrasted in the deterministic concepts of Relativity and the indeterministic concepts of Quantum Physics.

At the time of the formation of the Theory of Relativity, Quantum Theory had not yet reached the maturity of the present day; Einstein still worked on a unified field theory that would include a reconciled representation of quantum and gravitational phenomena, however, this time without success.

To date, such a theory called "Theory of Everything" continues not to exist, despite numerous proposals, the best known of which is the famous Superstring Theory.

1.2 WHO IS MOVING?

Let's take a nice walk, on a moonlit night. Are we sure we are the ones moving?

What does the moon think about this?

The impression of motion comes from seeing a series of objects behaving differently from my state of motion.

In a video game with a character or vehicle traversing a variety of landscapes, it is certainly not the latter that moves, but rather it is the pixels in the background that by moving transfer the feeling of movement to us.

Observing surrounding objects in a state of stillness with respect to my motion, it is difficult to imagine an inverse view of motion, because in the view of motion, the state of stillness obeys the rule that "the majority wins."

On a train, passengers and carriages move against the still state of the immense expanse of the surrounding landscapes.

A lone ship moves relative to the vast expanse of the sea.

So much so, that the ancients thought of the earth at rest and the sun in motion, because the observer placed on the earth observed a multiplicity of objects compared to the simple and solitary Sun.

But are we sure that this is not the case?

The truth is that it depends on points of view, that is, everything is relative.

I am taller than Gianni and less tall than Luca. When compared to Gianni I can be considered tall, vice versa when compared to Luca I am short.

If I consider an observation point sympathetic to the earth, all components of the universe move toward me.

If, on the other hand, I position myself in solidarity with the Sun, I will observe the Earth and other celestial elements always moving relative to my point of view (from the Sun).

What varies is only the geometric and mathematical representation of motion.

Under the heliocentric hypothesis, in describing the motion of the planets with respect to the Sun, we get the well-known elliptical orbits.

In geocentric theory, by describing the motion of the sun and planets with respect to the Earth, we roughly get "epicycles," as Ptolemy argued in about 100 CE.

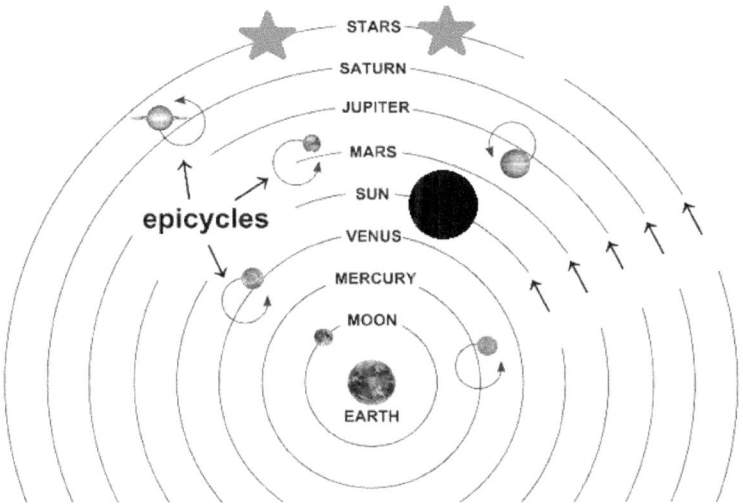

The explanation for the different ways of observing the same phenomenon lies in the non-existence of absolute motion and consequently the exclusive existence of only motions of a relative type.

However, the concept of relativity is not the result of Einstein's theory, but is an old issue considered as early as 1660 to be of fundamental importance in the representation of physical phenomena.

In fact, it is the scientist Galileo Galilei who is the true forerunner of relative systems.

In this regard, we quote a few lines from the great work written by Galileo Galilei about relative systems, *"Dialogue Over the Two Greatest Systems."*

Keep yourselves together with some friends in the largest room, which is under the cover of a large ship, and there let there be flies, butterflies and similar flying creatures; let there also be a large vase of water, and fish in it; and let some buckets be suspended on high, pouring water drop by drop into another vase with a narrow mouth, which is placed below: and standing still the vessel, observe diligently how those flying creatures with equal velocity go towards all parts of the room; the fishes will be seen to go noticing indifferently in all directions; the falling drops will all enter into the vessel underneath; and you, throwing to your friend some thing, you must not more gallantly throw it towards that part than towards this, when the distances are equal; and jumping you, as they say, at feet joined, equal spaces will pass towards all parts.

Observe that you will diligently have all these things, though no doubt there is no doubt that while the vassel stands still they should not so happen, make the ship move with as much velocity as you like; for (provided that the motion be uniform and not fluctuating here and there) you will not recognize the slightest mutation in all the named effects, nor by any of them can you understand whether the ship is walking or even standing still.

You jumping will pass in the planking the same spaces as before, nor, because the ship is moving very fast, will you make greater leaps toward the stern than toward the bow, although, in the time that you are in the air, the planking subjected to you will flow toward the side opposite to your leap; and throwing any thing to the mate, you will not with more force have to pull it, to reach him, if he be toward the bow and you toward the stern, than if you fuste situated for the opposite; the drops will fall as before into the lower vessel, without falling even one toward the stern, although, while the drop is in the air, the vessel is flowing many palms; the fish in their water will not with more effort notice towards the former than towards the succeeding part of the vessel, but with equal ease will come to the food placed on any place of the rim of the vessel; and finally the butterflies and flies will continue their flights indifferently towards all parts, nor will it ever happen that they will fall back towards the wall that concerns the stern, as if they were torn in keeping behind the swift course of the vessel, from which for a long time, holding themselves in the air, they will have been separated [...].

Note that the text is in Italian, but in the Italian language of the 600s.

Galileo, in his account, imagines being locked in the hold of a ship. In such a place he finds no difference in observing natural phenomena, compared to those who observe similar phenomena on land in a more obvious state of stillness.

Dropping something on the ground, or seeing a butterfly move, does not change whether we are locked inside a ship or in a room at home.

Similarly, the corresponding physical laws representing natural phenomenon also cannot have only privileged points of observation for which they are valid.

A physical law will have to be valid for all observers, even if they are moving.

It should be pointed out that Galileo's studies of relative motions are limited to systems (ship) in uniform rectilinear motion with respect to other systems (land), that is, without curving, accelerating or decelerating. We could also define uniform rectilinear motion as motion with constant velocity in both value and direction (must not decelerate) and direction (must not curve).

Before proceeding, it is important to clarify the concept of a reference system.

In physics, a reference system is given by a reference point coincident with the observer, from which I make my observations or measurements.

When we talk about the position or velocity (physical quantities) of a body, it is always necessary that they be expressed with respect to a "reference system," as the place where the observer is positioned.

Changing the reference system changes the values of observed physical quantities, such as position and velocity.

Think of a man sitting on a bench (observer). If this observes a person in the course of a jogging activity, with the aid of suitable instrumentation, he can easily measure the distance at which the jogger is placed and the speed of motion relative to himself.

Another person sitting in a circulating car will observe the sportsman at a different distance moving relative to himself at a different speed than observed by the man sitting on the bench.

In the first case, the reference system is the man sitting on the bench. This type of system we call it at rest, that is, stationary.

In the second case, the reference system is represented by the man sitting in the running car, so we are talking about a moving system.

The definition of motion or stillness of bodies is exclusively dependent on the existence of relative motion. Two people running at the same speed will never move apart and can continue to talk freely as if they were sitting still at a bench. They can be considered mutually at rest because their relative velocity is zero.

A reference system that is at rest or moving with constant velocity with no turns and no accelerations/decelerations (uniform rectilinear motion) is called an "inertial system."

A train moving on a rail at a constant speed represents an inertial system, otherwise a merry-go-round spinning with uniform circular motion represents a noninertial system since the speed although it does not change in value changes in direction.

Uniform rectilinear motion (Inertial reference frame)	Uniform circular motion (Non-inertial reference frame)

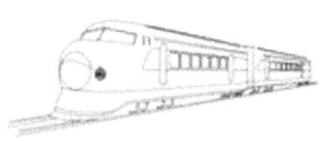 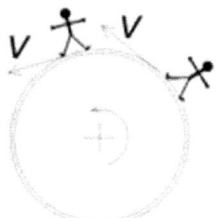

An exception is made for the Earth, which, although rotating, can be considered an inertial system to a good approximation.

Let's see why.

Let us consider the earth's rotational motion about its axis. Given a time equal to 24 hours to complete one full revolution and a value of the average Earth radius R equal to 6,371.00 km, we can calculate the average tangential velocity.

$$v = \frac{2\pi R}{24 * 60 * 60} = 463 \, m/s$$

The result obtained is not exactly a small value.

But a system is defined as inertial if it does not accelerate/decelerate or curve; speed has nothing to do with it.

Proceeding to the calculation of centripetal acceleration as a result of the Earth's rotation

$$a = \frac{v^2}{R} = \frac{1286^2}{6371000} = 0.03 \, m/s^2$$

and considering that the mean value of the Earth's gravitational acceleration is equal to $g = 9.81 \, m/s^2$, we quickly calculate the magnitude of centripetal acceleration with respect to gravitational acceleration, which turns out to be 0.3 percent of g. From the magnitude of this value, the approximation assumed in considering centripetal acceleration to be completely negligible for the observation of natural phenomena on the Earth's surface in the regime of an inertial reference system is evident.

An absolute reference system does not exist; every reference system is in motion or at rest always relative to other systems.

The inertial-type reference system is very important in physics in that the validity of physical laws are preserved in such systems, that is, the physical laws are "invariant" with respect to the passage of inertial reference systems.

The following example, again pertaining to relative motion, is very important for understanding different ways of representing a phenomenon, depending on the chosen reference system.

Examining a passenger in a train moving in uniform rectilinear motion (inertial system), given that he finds no difficulty in performing all common actions, it is possible to describe the phenomenon in two different ways:

1- The passenger moves along with the train, relative to the station with speed (v)

2- The person stopped at the station moves relative to the train-passenger by a speed (-v)

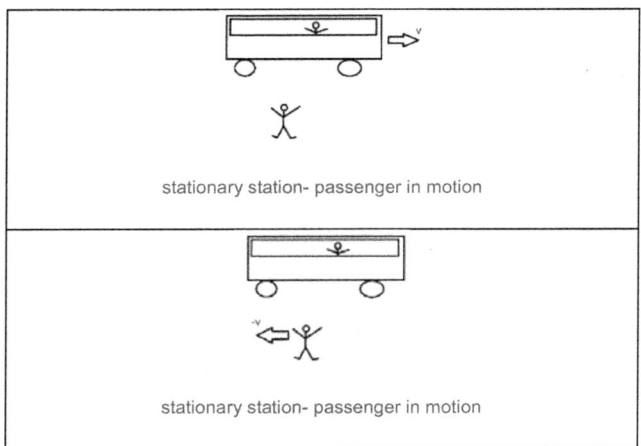

stationary station- passenger in motion

stationary station- passenger in motion

1.3 RELATIVITY

Galileo and, following, Newton studied in depth the relative motions of bodies although limited to cases of uniform rectilinear motion (inertial systems) and for velocity values contained to "human" values.

Under these conditions the corresponding physical laws were invariant, that is, they were valid as the reference system under consideration changed.

Things become more complicated the moment accelerated motions or motions at speeds comparable to those of light are considered.

In such a case one must abandon the study of relative motions according to Galileo and turn to the more complete and valid Theory of Relativity.

The Theory of Relativity is divided into two parts as chronologically presented by Einstein: Theory of Narrow (or Special) Relativity and Theory of General Relativity.

The Theory of Special Relativity introduces new concepts, based on the constancy of the speed of light (in a vacuum) and the abolition of the concept of absolute time hitherto used by classical physics. Time and metrics become elastic, causing time dilation and length contractions to be observed.

It is called "Narrow" because this theory considers only inertial (non-accelerated-non-curved) reference systems in the absence of gravity, analogous to Galileo's relativity but with velocities close to that of light.

It is evident how recourse to this theory becomes necessary only when velocity values of a certain magnitude come into play, that is, comparable to the speed of light, which is worth about 300,000 km/sec (to be exact, 299 792 458 m/s is the speed of light in a vacuum).

We will speak of high speeds if comparable to that of light and low speeds otherwise.

Just to get an idea of the order of magnitude, an airplane achieves speeds on the order of 1,000 kilometers per hour, equivalent to 0.331 kilometers per second, which compared with the speed of light appears decidedly low.

For low speeds Galileo's relativity continues to apply, which becomes a special case of the broader Theory of Special Relativity.

General Relativity also extends the theory to non-inertial (accelerated) reference systems and massive objects, thus including gravitational effects.

The result is a distortion of the four space-time dimensions.

People often make the mistake of associating the Theory of Relativity with the consequence that everything is Relative.

Nothing could be more wrong. The theory of relativity arose from the need to search for nonprivileged reference systems in order to represent phenomena with physical laws, which still remain valid in any reference system, even in motion.

In fact, the Theory rather than "of Relativity" should have been called the Theory on Invariants (physical laws invariant with different reference systems).

In other words, if I observe a phenomenon from our dear Earth, it must be represented by the same physical laws as an observer on Mars, without privileging either observer.

When this is not the case, the physical law surely hides unknown elements, and our ignorance leads us to resort to the existence of "apparent" elements or forces to prove its correctness.

If we are sitting in a car with totally tinted windows and moving forward at a steady speed, everything goes smoothly: I can play cards, read news on my smartphone, in perfect agreement with GALILEO's findings.

The moment the driver suddenly brakes, what happens? Me being in the car feels pushed forward, along with my playing cards, my smartphone ......and I wonder: what pushed me?

An observer sitting on the sidewalk, on the other hand, clearly sees a car braking and explains: the brake has locked the wheels, which then due to the frictional force generated by the adherence of the wheels with the road, the car has stopped along with all its passengers.

So we, confined in the car with tinted windows and oblivious to everything, hypothesize a mysterious force directed in the direction of motion, which propelled me forward along with all objects. I then invent an apparent force that is opposite and contrary to the deceleration force, so my physical model works perfectly again.

General Relativity was conceived precisely in the search for reference systems that are all equally valid, with no distinction between inertial and noninertial (accelerated) systems.

It is necessary to point out that Galileo's relativity, and as we shall later see also special (or restricted) relativity, are both valid only in inertial systems, that is, in systems at rest or in uniform rectilinear motion, which in addition to being a condition is a real limitation of representation of physical phenomena.

Galileo's relativity refers to motions at low speeds, while Einstein's special relativity refers to motions with speeds comparable to that of light.

Through the corresponding transformations, Lorentz's in the first case and Galileo's in the second case, it is possible to go from one reference system to another without the physical laws changing, always provided that the two reference systems are inertial.

Therefore, from an inertial system I can observe all phenomena with the physical laws always valid.

If, on the other hand, the observer is placed in an accelerated system, it is no longer possible to apply the Galileo/Lorentz transformations, and the physical laws are no longer the same, that is, they are no longer invariant.

It all depends on where the observer stands.

An observer at rest, will study the moving body in a circular motion (rotating), normally applying the laws of classical physics. The moving body will turn out to be subject to centripetal acceleration/force, which is that force required for the body in motion to curve and travel a circular trajectory.

An observer placed on the moving body, on the other hand, will no longer be able to apply the laws of classical physics except through the artifice of introducing a fictitious force: the centrifugal force, which is in the opposite direction to the centripetal force.

The above statement is more sympathetic when analyzing the behavior of an automobile on a curve.

Observing the motion from the dock, we will see the cornering car subject to a centripetal force, materialized by the friction of the wheels with the asphalt.

Observing the same phenomenon from inside the car, as a traveler, we will be stationary and therefore at rest with respect to the car. Since the car is subject to centripetal acceleration it happens that the car system is not inertial, so inside the car I can no longer apply the laws of classical mechanics. In fact if we drop an object in the car on a curve it will instead of falling vertically, curve outward due to a fictitious force called centrifugal force.

A schematic of the motion and forces involved follows.

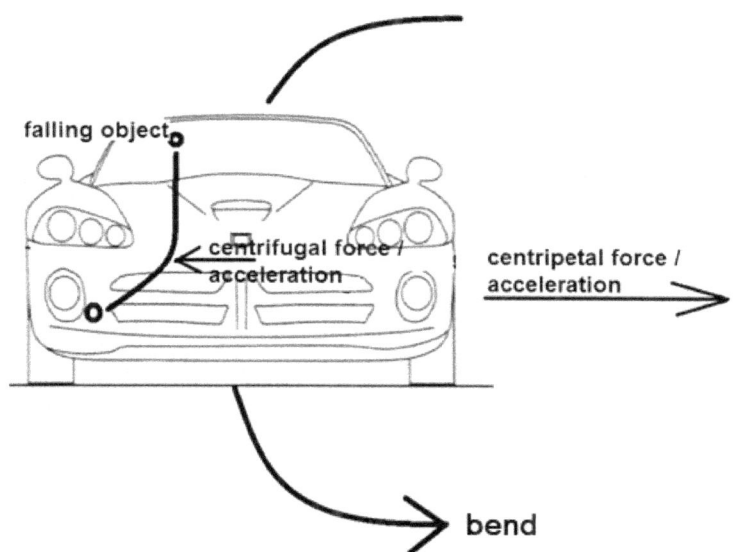

1.4 MOTIONS RELATED TO LOW SPEED (v<<c)

Let us first analyze relative motion in classical mechanics, with low velocities in play.

The mathematical exposition that follows, explains the existence of mathematical expressions suitable for transferring the representation of a physical formulation in different reference systems in relative motion (inertial type) with each other.

A physical law valid in a reference system in inertial motion, such as a train for example, can be transferred to a reference system at rest, at a station for example, simply through the use of appropriate formulas, called Galileo transforms.

Consider two reference systems, the first having O as its origin and the second having O' as its origin.

The reference system O is fixed, while O' and in relative motion relative to O, with a velocity v, in the direction of the X-axis (for simplicity, motion in only one direction is considered).

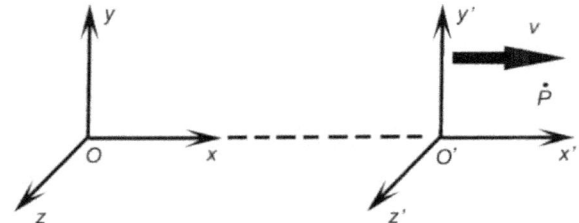

System O fixed, O' moving at speed v (P is stationary relative to O)

The point P is a fixed point, with respect to O, and will have coordinates in O of a fixed type and equal to (x,y,z), while the coordinates in O' will be of a variable type in the abscissa and equal to (x',y',z').

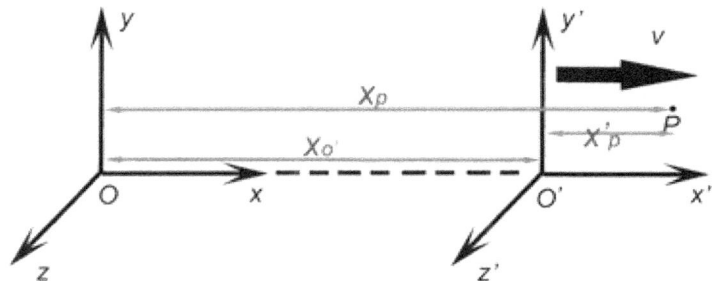

The coordinates of the point P in O are fixed, P being integral to O.

Differently in O', being in motion, its coordinate x'_p will be variable in time and in particular, its coordinate will be equal to :

$$(1)\ x'_p = x_p - x_{o'}$$

But, the coordinate $x_{o'}$ is nothing but the space traveled by the Reference System O' with velocity V, it follows that by simply applying the kinematic equation $v = \frac{s}{t}$ which in its inverse form becomes $s = v \cdot t$, we obtain:

$$(2)\ x_{o'} = v \cdot t$$

binding the two coordinate systems by making the due substitution of equation (2) into equation (1):

$$(3)\ x'_p = x_p - v \cdot t$$

Under this assumption, in view of the fact that motion is exclusive along the x-axis, it will result in y=y' and z=z', and the complete Galileo Transformations are written as:

$$(4)\ \begin{cases} x'_p = x_p - v \cdot t \\ y' = y \\ z' = z \\ t' = t \end{cases}$$

Just for completeness, Galileo's Transformations are also given for the more general case, with relative motion in the 3 dimensions x, y, and z:

$$(4a) \begin{cases} t' = t \\ x'_p = x_p - v_x \cdot t \\ y'_p = y_p - v_y \cdot t \\ z'_p = z_p - v_z \cdot t \end{cases}$$

Ultimately by means of the above Galileo transforms it is possible to use the physical laws on any other inertial reference system. In matrix form the above are written:

$$(4b) \quad \begin{bmatrix} t' \\ x'_p \\ y'_p \\ z'_p \end{bmatrix} = \begin{bmatrix} a_{11} & a_{12} & a_{13} & a_{14} \\ a_{21} & a_{11} & a_{23} & a_{24} \\ a_{31} & a_{32} & a_{33} & a_{34} \\ a_{41} & a_{42} & a_{43} & a_{44} \end{bmatrix} \begin{bmatrix} t \\ x_p \\ y_p \\ z_p \end{bmatrix}$$

and in a more schematic way

$$X' = A \cdot X$$

with **A** the transformation matrix, having some unit values, some null values, and some dependent on the components of the velocities

$$A = f(v_x ; v_y ; v_z)$$

As a special case, for the Galileo transformations in (4), that is, in the x-axis direction only, the matrix A becomes a simplified matrix:

$$(4c) \quad A = \begin{bmatrix} 1 & 0 & 0 & 0 \\ -v & 1 & 0 & 0 \\ 0 & 0 & 1 & 0 \\ 0 & 0 & 0 & 1 \end{bmatrix}$$

and consequently

$$\begin{bmatrix} t' \\ x'_p \\ y'_p \\ z'_p \end{bmatrix} = \begin{bmatrix} 1 & 0 & 0 & 0 \\ -v & 1 & 0 & 0 \\ 0 & 0 & 1 & 0 \\ 0 & 0 & 0 & 1 \end{bmatrix} \begin{bmatrix} t \\ x_p \\ y_p \\ z_p \end{bmatrix}$$

While in the more general case of (4a)

$$(4d) \qquad A = \begin{bmatrix} 1 & 0 & 0 & 0 \\ -v_x & 1 & 0 & 0 \\ -v_y & 0 & 1 & 0 \\ -v_z & 0 & 0 & 1 \end{bmatrix}$$

we will have

$$\begin{bmatrix} t' \\ x'_p \\ y'_p \\ z'_p \end{bmatrix} = \begin{bmatrix} 1 & 0 & 0 & 0 \\ -v_x & 1 & 0 & 0 \\ -v_y & 0 & 1 & 0 \\ -v_z & 0 & 0 & 1 \end{bmatrix} \begin{bmatrix} t \\ x_p \\ y_p \\ z_p \end{bmatrix}$$

The same relationships can be inverted from the reference system O' to the reference system O, using the matrix A⁻¹ inverse

$$X = A^{-1} \cdot X'$$

Where in the more general form of motion in three dimensions

$$A^{-1} = \begin{bmatrix} 1 & 0 & 0 & 0 \\ v_x & 1 & 0 & 0 \\ v_y & 0 & 1 & 0 \\ v_z & 0 & 0 & 1 \end{bmatrix}$$

Just to become familiar with the mathematical language used by Einstein, the same equations can be written by adopting precisely "Einstein's notation," also called "Einstein's convention in summations."

Said convention was introduced by Albert Einstein himself to make some of the equations of general relativity more concise, but without their having any physical meaning, being merely a more useful method of writing in mathematical formalism.

By indicating the spatial and temporal coordinates with such a formalism, we will have

$$x^1 = t \; ; \; x^2 = x \; ; \; x^3 = y \; ; \; x^4 = z$$

and (4b) is written

$$(4e) \quad x'^v = \Lambda^v_\mu \, x^\mu$$

denoting a summation with $v = 1,2,3,4$ as $\mu = 1,2,3,4$ varies, in Euclidean space, with $a_{v\mu}$ which components of the matrix Λ^v_μ

$$x'^v = \sum_{\mu=1}^{4} \left(a_{v\mu} \cdot x^\mu \right)$$

The preceding at $v = 1$ becomes

$$x'^1 = \sum_{\mu=1}^{4} \left(a_{1\mu} \cdot x^\mu \right) = a_{11} \cdot x^1 + a_{12} \cdot x^2 + a_{13} \cdot x^3 + a_{14} \cdot x^4$$

continuing, with $v = 2$, $v = 3$ and v=4

$$x'^2 = \sum_{\mu=1}^{4} \left(a_{2\mu} \cdot x^\mu \right) = a_{21} \cdot x^1 + a_{22} \cdot x^2 + a_{23} \cdot x^3 + a_{24} \cdot x^4$$

$$x'^3 = \sum_{\mu=1}^{4} \left(a_{3\mu} \cdot x^\mu \right) = a_{31} \cdot x^1 + a_{32} \cdot x^2 + a_{33} \cdot x^3 + a_{34} \cdot x^4$$

$$x'^4 = \sum_{\mu=1}^{4} \left(a_{4\mu} \cdot x^\mu \right) = a_{41} \cdot x^1 + a_{42} \cdot x^2 + a_{43} \cdot x^3 + a_{44} \cdot x^4$$

All the above equations are equivalent to the more general case in (4b).

For a Galilean transformation with motion in the three dimensions the components $a_{v\mu}$ take on the values given in the A matrix as shown in (4d), and for a Galilean transformation with motion in the x-axis dimensions only the components take on the values represented in the A matrix of equation (4c).

1.5 GALILEO TRANSFORMATIONS

To better understand the meaning of the result obtained in the previous paragraph, it is useful to verify how Newton's Second Law is invariant with respect to an inertial reference system, that is, how the corresponding Physics law remains unchanged as the reference system under consideration changes, again under inertial conditions.

Newton's formulation in the reference system at rest O is:

$$(5)\ F = m \cdot a$$

For simplicity we assume, a motion exclusively along the x-axis and a force acting along the same direction.

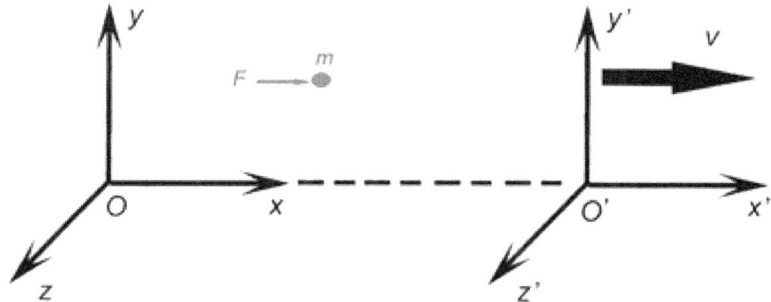

(5) can be written in differential form $F = m \cdot \dfrac{d^2 x}{dt^2}$

Now we move to the reference system O', in relative motion with respect to O, by applying Galileo's Transformations (4), making explicit and substituting in the variable x

$$F = m \cdot \frac{d^2(x' + v \cdot t)}{dt^2} = m \cdot \frac{d^2(x')}{dt^2} + m \cdot \frac{d^2(v \cdot t)}{dt^2}$$

$$= m \cdot \frac{d^2(x')}{dt^2} + m \cdot v \cdot \frac{d^2 t}{dt^2}$$

Given that under low-speed motion time is universal, we can

impose $\frac{d^2 t}{dt^2} = 0$, thus obtaining

$$F = m \cdot \frac{d^2(x')}{dt^2}$$

and again, remembering that the second derivative of position
with respect to time represents acceleration

$$F = m \cdot a'$$

The equation obtained turns out to be of the same form as (5),
and thus we can admit that Newton's second law is "invariant" in
the transition between inertial systems.

1.6 GALILEO TRANSFORMATIONS AND LIGHT

Let us now analyze what happens when observing a light signal (understood as rays or as a set of Photons), from the point of view of two reference systems, one at rest and one in inertial motion.

In this hypothesis, it will be shown that Galileo's transformation laws no longer hold true for the light signal, which travels at the speed of light, unlike the material point moving at a reduced speed.

Galileo's transformations have limits of applicability, precisely, so it will have to be necessary to search for additional laws of transformations that are always valid, in all cases.

Let us carry out the analytical demonstration.

We apply Galileo's transformations and consider the speed of light to be constant regardless of reference systems.

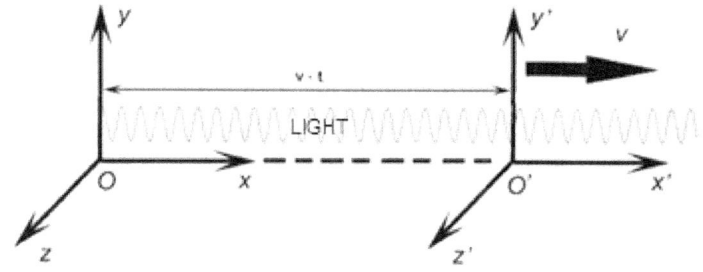

The space traveled by the light signal in the reference system O is equal to:

$$(6)\ x = c \cdot t$$

Applying Galileo's transformations (3)

$$(7)\ x' = x - v \cdot t$$

Substituting (6) into (7).

$$(8)\ x' = c \cdot t - v \cdot t$$

Since the speed of light (c) is constant in each inertial reference system, in the reference system O' the light signal travels a space equal to:

$$(9)\ x' = c \cdot t$$

Substituting (9) into (8).

$$(8)\ c \cdot t = c \cdot t - v \cdot t$$

Simplifying t to both members for t >0

$$(9)\ c = c - v$$

That represents an impossible outcome.

Having obtained on the basis of a hypothesis, an impossible result, consequently the hypothesis initially formulated also becomes false and it can be deduced that Galileo's transformations are no longer applicable when dealing with particles traveling at the speed of light (Photons).

1.7 SIMULTANEITY OF EVENTS IN RELATIVE MOTIONS

Position, explicated through the identification of the three coordinates x,y and z, assumes a unique and universal type of time, constantly flowing in one direction.

In relativity, in consideration of the different possible times, it will be necessary to abandon the primitive thinking of position (x,y,z) and replace it with the innovative concept of "event."

An event is representative not only of location but also of the time in which it occurs (x,y,z,t).

Two events are simultaneous when they happen at the same time, more or less !!!

Let us begin to explore the concept of Simultaneity by considering "sound."

Sound needs an environment, called a medium (air, water, iron, etc.), to propagate. An alarm clock placed inside a vacuum glass bell will not emit any sound.

The rate of propagation is variable with temperature and the type of environment.

In air at a temperature of 0°C its value is 331 m/s, in iron it is about 5,000 m/s, all still modest values compared to the already famous speed of light of 299,792,458 m/s (in vacuum).

Einstein jokingly stated: to hear a lost word again, one need only run at a speed faster than the speed of sound to catch it.

Let us carry out some practical examples, considering a Train, a Siren and two People.

The event is represented by the sound of the siren. We will consider here the sound of the siren as a single emission and not continuously like ambulances for instance. The experiments are performed under ideal conditions, neglecting air-train friction. Let's examine what happens in the different possible combinations of Train stopped-moving, passengers on board (windows closed)-on the ground, and indoor-outdoor siren.

1) Train stopped - Interior siren - Interior passengers placed at equal distance from the siren

In this situation, the event is **simultaneous** for the two observers; the sound reaches the two observers at the same time.

2) Train stopped - Interior siren - Exterior passengers placed at equal distance from the siren

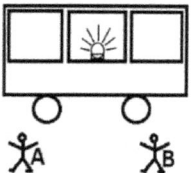

The event is **simultaneous**, like the previous case. The sound reaches the two observers simultaneously.

3) Train in motion - Interior siren - Interior passengers placed at equal distance from the siren

In this situation, the event is **simultaneous** for the two observers, since the medium (the air inside the train....with windows closed) is integral to the train and therefore the speed of sound propagation is the same in the direction of motion and in the opposite direction.

4) Train in motion - Interior siren - Exterior passengers placed at equal distance from the siren

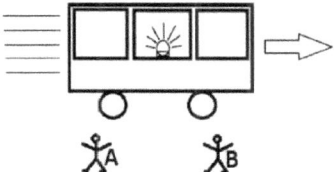

The event is **NOT simultaneous**; the sound reaches observer B first and then observer A. The sound moves faster in the direction of motion with increased speed just by the speed of the train and thus the source of the sound. Observer A sees the sound moving at a speed equal to $v_{sound} - v_{train}$. Observer B sees the sound moving at a speed equal to $v_{sound} + v_{train}$.

5) Train in motion - Exterior siren - Interior passengers placed at equal distance from the siren

The event is **NOT simultaneous**; the sound reaches first observer A and then observer B. Observer A, in motion, goes to meet the sound propagating in the medium, at rest, placed outside, is equivalent to saying that the sound moves faster in the direction opposite to the motion with increased speed just by the speed of the train. Observer A sees sound moving at a velocity equal to $v_{sound} + v_{train}$. Observer B sees sound moving at a velocity equal to $v_{sound} - v_{train}$. What matters is the relative velocity, in this case observer A approaches the source and observer B moves away from the source. It is clear how this case is identical to case 4) with observers A-B having reversed priority.

6) Train in motion - External siren - External passengers placed at equal distance from the siren

In this situation the event is **simultaneous** for the two observers. The medium (the air outside the train) is integral to the reference of the two observers (at rest), and therefore the speed of sound

propagation is the same in the direction of motion and in the opposite direction.

The physical laws representing the previous phenomena, which can be traced back to calculations of velocity composition, not comparable to the speed of light, are clearly invariant to Galileo's Transformations.

1.8 REASONS RELATED TO HIGH SPEED (v $\simeq$ c)

So far we have seen how relative motions at low velocities are quite simple to solve because of their invariance to the simpler Galileo Transforms.

Problems about relative motions arise by considering electromagnetic phenomena, where electromagnetic waves like light itself travel at the limiting speed "c" (speed of light).

James Clerk Maxwell with the unification theory of electromagnetism, elegantly expressed by the formulation of the famous equations, in addition to introducing a constant equal to c equal to the speed of light in a vacuum, completely solves the link between electric field and magnetic field. With Maxwell, electricity and magnetism become two sides of the same coin, as does the behavior of light and an electromagnetic wave.

Recalling that usually the letter E is used to denote the Electric Field and the letter B the Magnetic Field, the above equations are given for knowledge only, without utility for the topics that follow.

Name	Differential form	Integral form
Gauss's Law	$\nabla \cdot \mathbf{D} = \rho_f$	$\oiint_{\partial V} \mathbf{D} \cdot d\mathbf{A} = Q_f(V)$
Gauss's Law for magnetism	$\nabla \cdot \mathbf{B} = 0$	$\oiint_{\partial V} \mathbf{B} \cdot d\mathbf{A} = 0$
Faraday's law of induction	$\nabla \times \mathbf{E} = -\dfrac{\partial \mathbf{B}}{\partial t}$	$\oint_{\partial S} \mathbf{E} \cdot d\mathbf{l} = -\dfrac{\partial \Phi_{B,S}}{\partial t}$
Ampere's law (with Maxwell's correction)	$\nabla \times \mathbf{H} = \mathbf{J}_f + \dfrac{\partial \mathbf{D}}{\partial t}$	$\oint_{\partial S} \mathbf{H} \cdot d\mathbf{l} = I_{f,S} + \dfrac{\partial \Phi_{D,S}}{\partial t}$

The above equations turn out to be "non-invariant" with respect to reference systems in relative inertial motion, that is, they do not respect invariance by means of the Galileo transforms.

Here is immediately created the split.

On the one hand we have classical mechanics that for systems in inertial motion, through Galileo's transformations, the corresponding physical laws are invariant, and on the other hand we have electromagnetism that for systems in motion Maxwell's equations are not invariant.

So something doesn't add up.

It appears that Electromagnetic phenomena are represented by equations that are valid exclusively in a kind of privileged system, in total contrast to Galileo's relativistic principle.

At the time, an attempt was made to give an explanation for what was happening by introducing the Ether. Ether was supposed to be "a medium" that allowed the diffusion of LIGHT in the same way that the medium "air" allowed the diffusion of sound.

In this regard, several experiments were performed to prove the variability of the speed of light as the speed of reference systems changed.

In 1881 Albert Abraham Michelson, in order to verify the existence of the aether, decided to measure the speed of light in different directions using an instrument he designed, Michelson's interferometer. The interferometer consists of a light source that directed onto an inclined semi-reflecting mirror (half reflecting and half letting it go on) allows a beam of light to be split into two beams that travel following perpendicular paths and are then converged again, through other reflecting mirrors, onto a screen forming an interference figure. The interference figure is obtained by superposition of the maxima and minima of the light waves, creating light bands and dark bands.

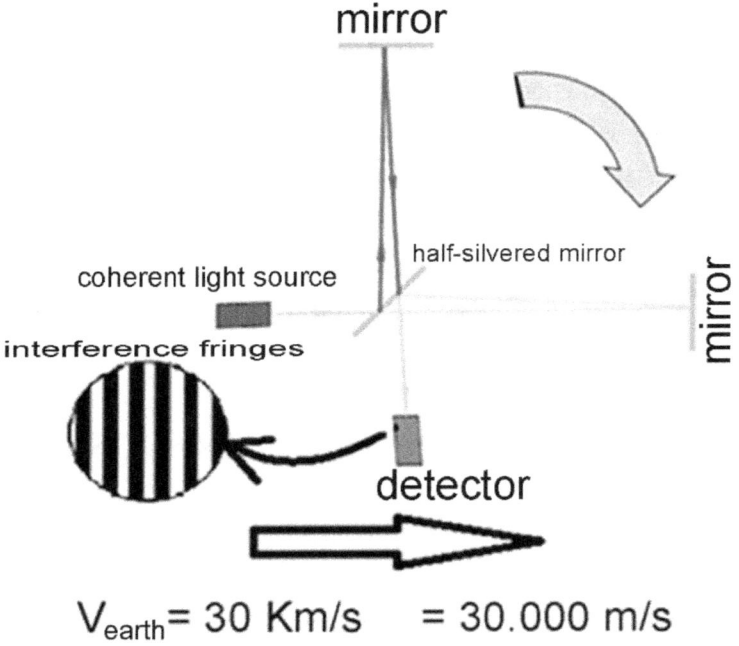

$$V_{earth} = 30 \text{ Km/s} = 30.000 \text{ m/s}$$

Michelson experiment schematization
(light beams do not just overlap for
a more explicit graphical representation)

The Earth, because of its motion within the solar system (motion of revolution), should have encountered an ether "wind" at 30 km/s.

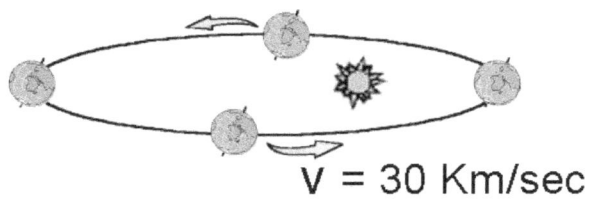

$$V = 30 \text{ Km/sec}$$

Anything immersed in the ether would be affected by the wind, including light.

As the instrument is rotated, during the same rotation, a phase shift of arrival at the detector of the two separate beams of light should occur, in relation to the congruence of the direction and direction of light travel with the direction and direction of Earth's rotation. As a consequence of the arrival fse difference, a slip of the interference bangs materializes as the apparatus rotates with respect to the direction of the ether wind.

Using this experimental device, Michelson made a number of measurements and detected no expected shift in the interference bangs. This result was also confirmed later in 1887, when in collaboration with Edward Morley, he performed a new experiment with higher precision.

The result obtained although it established the nonexistence of ether wind with the speed of light shown to be constant in moving systems, it complicated the search for a soluzone of the invariance problem of Maxwell's equations.

This experiment, however, did not influence Einstein's studies in any way, who according to him, was already firmly convinced of the nonexistence of absolute motion and the phantom Aether.

The physicist-mathematician Hendrik Antoon Lorentz intervened in this regard, and through purely mathematical contrivances he found new laws of transformations that were able to make Maxwell's equations "invariant."

Lorentz transforms also become valid for inertial motions at high velocities, thus retiring the Galileo transforms.

The problem seemed, then, to be definitively solved, but so it was not because the Lorenzt transformations remained a mathematical artifice without any physical meaning.

Let us analyze such transformation laws analytically.

Considering, for simplicity, a motion along the x-axis, the Lorentz transforms are as follows:

$$
\begin{cases}
x' = \dfrac{x - v\,t}{\sqrt{1 - \dfrac{v^2}{c^2}}} \\[4ex]
y' = y \\[1ex]
z' = z \\[2ex]
t' = \dfrac{t - \dfrac{v}{c^2}\,x}{\sqrt{1 - \dfrac{v^2}{c^2}}}
\end{cases}
$$

With

x', y', z' coordinates of the reference system in relative motion

x, y, z coordinates of the reference system at relative rest

t = time in the reference system at relative rest

t' = time in the reference system in relative motion

v = velocity of the reference system in relative motion

c = speed of light

More simply, equations can also be written:

$$
\begin{cases}
t' = \gamma \left(t - \frac{v}{c^2} x \right) \\
x' = \gamma \left(x - vt \right) \\
y' = y \\
z' = z
\end{cases}
\qquad
\gamma = \dfrac{1}{\sqrt{1 - \dfrac{v^2}{c^2}}}
$$

For those who would like to investigate, as with Galilean transformations, the matrix form of Lorentz transformations, introducing the quadrivector

$$\begin{bmatrix} ct \\ x \\ y \\ z \end{bmatrix}$$

Equations can be written in:

$$\begin{bmatrix} ct' \\ x' \\ y' \\ z' \end{bmatrix} = \begin{bmatrix} a_{11} & a_{12} & a_{13} & a_{14} \\ a_{21} & a_{11} & a_{23} & a_{24} \\ a_{31} & a_{32} & a_{33} & a_{34} \\ a_{41} & a_{42} & a_{43} & a_{44} \end{bmatrix} \begin{bmatrix} ct \\ x \\ y \\ z \end{bmatrix}$$

or

$$X' = \Lambda \cdot X$$

With Λ the transformation matrix, where some of its values are unitary, some are zero, and some are a function of the components of the velocities

$$\Lambda = f(v_x ; v_y ; v_z)$$

Particularly in the case of motion in the x-axis direction only, the matrix Λ becomes a simplified matrix:

$$\Lambda = \begin{bmatrix} \gamma & -\gamma \dfrac{v}{c} & 0 & 0 \\ -\gamma \dfrac{v}{c} & \gamma & 0 & 0 \\ 0 & 0 & 1 & 0 \\ 0 & 0 & 0 & 1 \end{bmatrix}$$

and consequently

$$\begin{bmatrix} c\,t' \\ x'_p \\ y'_p \\ z'_p \end{bmatrix} = \begin{bmatrix} \gamma & -\gamma\dfrac{v}{c} & 0 & 0 \\ -\gamma\dfrac{v}{c} & \gamma & 0 & 0 \\ 0 & 0 & 1 & 0 \\ 0 & 0 & 0 & 1 \end{bmatrix} \begin{bmatrix} c\,t \\ x_p \\ y_p \\ z_p \end{bmatrix}$$

The same relationships can be inverted from the reference system O' to the reference system O, using the matrix Λ⁻¹ inverse

$$X = \Lambda^{-1} \cdot X'$$

And again, the equations of the Lorentz transformations can, like Galileo's previous ones, be written adopting Einstein's notation

$$x'^{\nu} = \Lambda^{\nu}_{\mu}\, x^{\mu}$$

which in inverse form becomes

$$x^{\nu} = \Lambda^{-1}{}^{\nu}_{\mu}\, x'^{\mu}$$

With the same meanings as the elements in (4e) and with the coordinates equal to

$$x^1 = c \cdot t \; ; \; x^2 = x \; ; \; x^3 = y \; ; \; x^4 = z$$

Note that instead of considering the time coordinate t, a spatial equivalent coordinate is used $c \cdot t$,having multiplied a time quantity by a velocity quantity.

This artifice allows a time coordinate to be transformed into a metric coordinate (velocity x time = space).

As we shall see below, the four coordinates thus structured (quadrivector) are the basis for the establishment of a Minkowski spacetime (M4).

1.9 SIMULTANEITY IN THE "LIGHT" EVENT

Let us now try performing "mental experiments" of simultaneity, considering instead of a sound source, as in the previous paragraphs, a light source (LIGHT).

First of all, LIGHT, unlike SOUND, propagates without means, even in a vacuum, and its speed is extremely great and constant. Let's analyze the moving train experiment.

Consider the usual two observers inside the moving train and two additional observers on the ground. The event is represented by the emission of a light signal that reaches observers A and B inside the train.

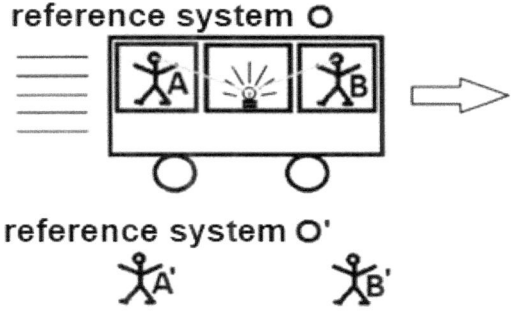

Let us place ourselves in O', a reference system at relative rest, and observe the simultaneity conditions.

Observers A' and B' will SIMULTANEOUSLY observe the event by being at rest.

To observers A and B, however, the light signal will first reach A and then B, as A approaches the light signal while B moves away. For A and B, the event will be nonsimultaneous.

By reciprocity, let us place ourselves in O and consider the wagon as the reference system, such that the wagon is at relative rest and the observers A' and B' are in relative motion, integral to the system O'.

system O

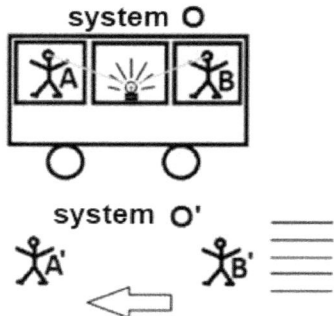

Under this assumption, observers A and B will SIMULTANEOUSLY observe the event.

While observers A' and B' will get the light signal first in B' and then in A', as B' approaches the light signal while A' moves away.

Now to the question of whether the event is simultaneous or not we are forced to answer that the same event can be SIMULTANEOUS or NOT, depending on the reference system considered.

We get an identity crisis of the absolute concept of simultaneity, and we are forced to specify with respect to which reference system (at rest or in motion) we intend to evaluate the event.

CHAPTER 2
THEORY OF SPECIAL RELATIVITY

2.1 THE POSTULATES OF SPECIAL RELATIVITY

The resolution of the problem, seen in the previous paragraph concerning relativity with electromagnetic phenomena (light phenomena), is finally solved, within a few months, by Albert Einstein in 1905, through his insight to move away from any prejudice about the belief of considering time as "absolute," by introducing a new and innovative concept of "relative time."

Already St. Augustine (400 A.D.) one of the greatest philosophers of the first millennium, to the question "What then is time?" answered "When no one asks me I know, but if someone asks me and I want to explain it to him, I don't know."

In modern times, John A. Wheeler, a distinguished American physicist, following the theory of relativity, defined time as "the way nature prevents things from happening all at once."

Ultimately, each reference system has its own time that implies "relative simultaneity": Observers in relative motion measure different times that depend on the speeds with which they move, resulting in perceiving the same event at different times.

After all, Einstein uses mathematical relations and experiments already known from other physicist-mathematicians, but adds, as the fruit of his own "genius," the idea of abandoning the hinged concept of absolute time. Einstein introduces with relativity the irrational and new concept of relative time.

All here, the theory of Narrow (or Special) Relativity was born with the two postulates:

- All laws of Physics (mechanics- **electromagnetism**) are the same in all **inertial reference systems** (in uniform rectilinear motion with respect to other systems);
- In inertial systems, light propagates in a **vacuum** at constant speed, regardless of the state of **motion** of the source or observer.

2.2 TIME DILATION

Introducing different times for reference systems in relative motion leads to an important and innovative consequence: time dilation.

An observer at rest will observe a dilated time of the actions of another observer in motion, at speeds close to light.

If an observer at rest, for example, measures a time of 10 minutes for an event taking place on a moving reference system, at high speeds, for the observer integral to the latter, the elapsed time will be only 5 minutes.

Let's now try to do a few little calculations, considering the usual two reference systems in relative inertial motion with each other or rather one system at rest and the other in uniform rectilinear motion at high speed comparable to c.

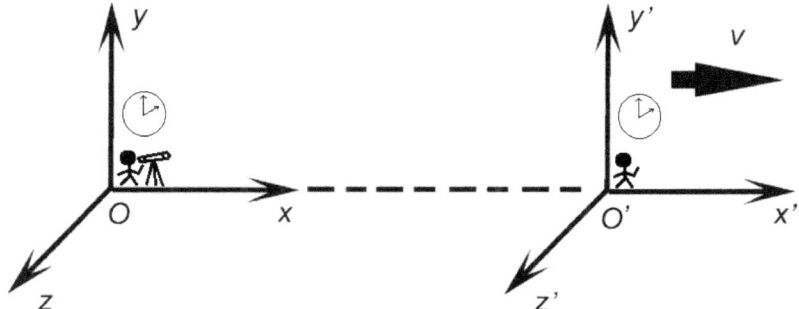

System O will measure a time t, while system O' will measure a time t'.

We analyze the time variable using the Lorentz relation that links the times of the two reference systems:

$$(10)\ t' = \frac{t - \frac{v \cdot x}{c^2}}{\sqrt{1 - \frac{v^2}{c^2}}}$$

By posing $\beta = \frac{v}{c}$ (10) becomes

$$(11)\ t' = \frac{t - \frac{v \cdot x}{c^2}}{\sqrt{1 - \beta^2}}$$

Let us now consider two consecutive instants: instant 1 and instant 2 and rewrite (11) for both instants:

$$(12)\ t_1' = \frac{t_1 - \frac{v \cdot x_1}{c^2}}{\sqrt{1 - \beta^2}}$$

$$(13)\ t_2' = \frac{t_2 - \frac{v \cdot x_2}{c^2}}{\sqrt{1 - \beta^2}}$$

Subtracting (13) with (12) and substituting ($dt = t - t_{21}$) and ($dt' = t'_2 - t'_1$), we get:

$$(14)\ dt' = \frac{dt - \frac{v \cdot dx}{c^2}}{\sqrt{1 - \beta^2}}$$

That was equivalent to writing (11) in differential form.

Now instead we consider the reference system O and the space traveled in time dt, which is equal to:

$$(15)\ dx = v \cdot dt$$

Substituting (15) into (14).

$$dt' = \frac{dt - \frac{v^2 \cdot dt}{c^2}}{\sqrt{1 - \beta^2}}$$

highlighting *dt* in numerator

$$dt' = \frac{dt(1 - \frac{v^2}{c^2})}{\sqrt{1 - \beta^2}}$$

Remembering that we had placed $\beta = \frac{v}{c}$

$$dt' = \frac{dt(1 - \beta^2)}{\sqrt{1 - \beta^2}}$$

Streamlining

$$dt' = \frac{dt(1 - \beta^2)}{\sqrt{1 - \beta^2}} \cdot \frac{\sqrt{1 - \beta^2}}{\sqrt{1 - \beta^2}}$$

$$(16)\ dt' = dt \cdot \sqrt{1 - \beta^2}$$

Writing (16) in the inverse form in *dt*

$$(17)\ dt = \frac{dt'}{\sqrt{1 - \beta^2}}$$

and placing

$$\gamma = \frac{1}{\sqrt{1 - \beta^2}}$$

The (17) becomes

$$(18)\ dt = \gamma\, dt'$$

The formulation of the variable γ leads to an important consideration: the permissible velocities are always less than that of light, since for higher velocities the denominator of (10) becomes impossible to extract, as the root of a negative number. Since $\beta < 1$ since $v < c,$ it follows that the factor γ is always greater than unity $(\gamma > 1)$ and therefore, relative to an event occurring in O', the time measured by O is greater than the time measured in O', or rather the observer in O measures a dilated time relative to what is measured by O'.

2.3 TWINS PARADOX

A famous consequence of the possibility of time dilation allows us to enunciate a famous paradox.

Two twins, living on Earth, one fine day decide to separate.

The first (A) stays on Earth and the second (B) moves away by boarding a spacecraft moving at near-light speed.

The twin left on the ground observes a dilated time (runs slower) than the twin's life in the spacecraft, that is, longer than the twin on the spacecraft perceives and measures.

Introducing some numbers, the twin on the ground could measure twice as long as the twin in motion, so if the twin in motion measures and 30 years elapse for him, the twin on the ground measures an elapsed time equal to 60 years. Consequently upon the return of the twin in motion, he will find his twin left on the ground older.

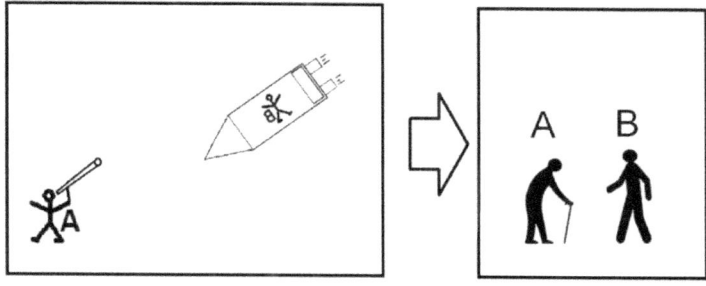

What has just been described bears the name of the famous thought experiment by the name PARADOX OF THE JEWS.

Let us now see why it is a paradox.

By reciprocity, if we consider the reference system in solidarity with the moving twin, the latter will appear to be at rest.

As a result, the twin on earth will appear in motion, through a simple play of relative velocities.

The paradox poses the question that in the first case, the twin A on the ground when reunited with the twin B who has traveled will be older than the latter. This is because A observes a dilated time of B

Conversely, in the second case, considering for the sake of reciprocity A in motion and B at rest, we will have that B will be older than its twin A.

The question is, who is actually younger, A or B?

The explanation follows.

In order for the theory of Special Relativity to apply, it is necessary for the reference system to be inertial (at rest or in uniform rectilinear motion)

The reference system in A, is at rest and therefore the Theory of Special Relativity applies.

The reference system in B, in view of the fact that to reach the speed of light it must accelerate and to return and meet with A it must decelerate and curve in the course of the reversal, is not an inertial system and therefore the above theory of Special Relativity cannot find application.

In this interpretation, the PARADOX is generated by the misapplication of a theory

A further interpretation of the PARADOX is the condition that the twins once B has left do not meet again.

Under such assumptions, the theory of Special Relativity remains valid, provided that twin B behaves as an inertial system (in uniform rectilinear motion without making deflections or decelerations) and if, in any case, the acceleration of B required to reach the high speeds is neglected.

The principle of reciprocity also remains valid: twin A will see (as long as he can) younger B in the same way that B will observe younger A, but without both of us having the opportunity to meet and verify it.

The diversity of time measurement is solely related to the location of the event you want to observe, in relation to your speed of movement.

What is relative is the measurement of time with reference to the concept of SIMULTANEITY, while in reality each of the twins lives his or her own life, certainly not simultaneously in the rational sense of the term.

Within the scope of the universe, every living and non-living being lives its life to the passage of its own time, which to an outside observer might certainly appear different.

The concept of time, as it is normally understood, with Einstein's relativity changes meaning when we are faced with high speeds or, as we will see below with General Relativity, when dealing with very large objects (stars, black holes, etc.).

Recent experiments with aircraft in flight equipped with high-precision (cesium atomic) on-board clocks have found that time dilation, as predicted by the theory of Special Relativity, is perfectly true and in agreement with theoretical results. The clock traveling on the plane turns out to be slowed down compared to a synchronized clock placed on the ground, at the time of comparison upon its return.

If the velocities involved are very low compared with the speed of light, the factor β tends to zero

$$\beta = \frac{v}{c} = 0$$

and then

$$\gamma = \frac{1}{\sqrt{1 - \beta^2}} = 1$$

(18) becomes $dt = dt'$, that is, time returns to being absolute and unique.

2.4 PROPER TIME

The characteristic of relativity to have different times for different reference systems leads to the definition of "proper time," denoted by the Greek letter τ (tau).

Proper time is the time measured in the reference system integral with the observed event or phenomenon. And it is always the shortest time.

In the twin case, A observes B's event and measures a dilated time of it, so the proper time τ is the lesser of the two and is precisely the measured time where the event you want to observe (measured by B) takes place.

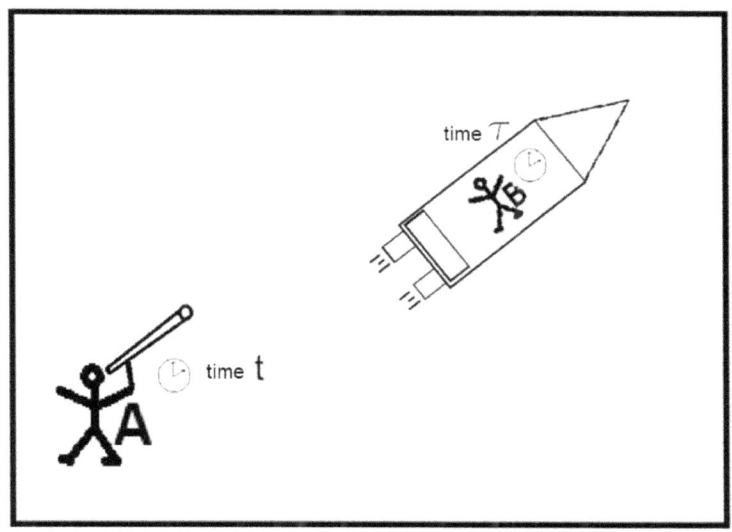

Having introduced proper time, (18) can be written as.

$$(18)\ dt = \gamma\ d\tau$$

With > 1 , dt is always greater than $d\tau$.

2.5 EXPERIMENTAL FINDINGS

To perform particle experiments at high velocities, it is necessary to use particle accelerators, resulting in great energy expenditure.

Another, simpler way to perform experiments with high-speed particles is to observe cosmic rays (particles from the cosmos).

Cosmic rays are formed at an altitude of about 10 km and, traveling at a speed almost equal to that of light, penetrate the upper layers of the atmosphere, generate pions, which in turn decay into muons and neutrinos and finally reach the Earth's surface.

The correct interpretation of their observation is obtained by considering the relativistic concept on time dilation.

In particular, Muon or Meson mu (particles consisting of a Quark and an anti-Quark) decay (transform) into other elementary particles (electron-antineutrino-neutrino, positron-antineutrino-neutrino) in an average time (half-life) of $2,15 \ 10^{-6} \ secondi$.

The speed of mesons is relativistic type of about *0.98967* times the speed of light *c*.

Performing a nonrelativistic calculation, it is found that the Mesons before decaying should take a path equal to

$$s = v \cdot t = \ 638 \ \text{mt}$$

They should not, therefore, be observed at ground level as they are in reality.

Theoretical confirmation comes to us by applying the theory of special relativity, because of the characteristic possibility of time dilation.

Knowing that mesons have a half-life "proper τ" equal to $2{,}15\ 10^{-6}\ secondi$, the theory of relativity imposes for an observer placed on the earth's surface a measurement of a different and dilated time dt.

The latter time dt can be derived known as time dt' (also referred to as $d\tau$) using (18), after calculating the corresponding value of γ

$$\gamma = \frac{1}{\sqrt{1 - \beta^2}} = 6.975$$

$$\beta = \frac{v}{c} = 0.98967$$

$$dt = 6.975 \cdot 2.15\ 10^{-5} = 1.5\ 10^{-5}$$

At this point it is possible to calculate the actual space traveled by Muons before they decay (turn into something else)

$$s = v \cdot dt = 4452\ m$$

Here is demonstrated through the use of special relativity the observation of Muons at altitudes compatible with experimental observations.

2.6 LENGTH CONTRACTION

In relativity it happens that like times, lengths have different measurements depending on the chosen reference system.

A stationary observer observes moving objects contracted with respect to the length measured by the moving observer integral to the object.

Let us consider, as usual, two reference systems in relative motion, for simplicity, along the x-axis only.

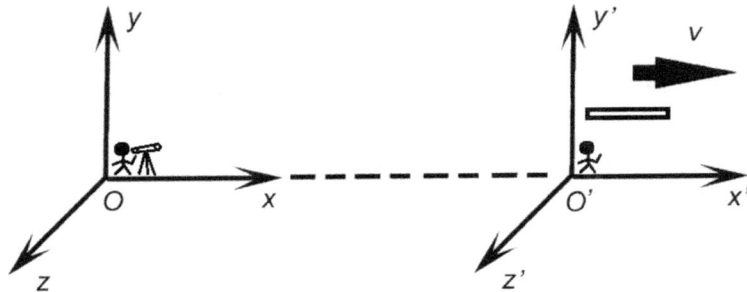

The reference system O is fixed, while the reference system O' moves in uniform rectilinear motion, together with the observer and the ruler of length L' (measured in O').

Using Lorentz transformations in differential form:

$$(19) \ dx' = \frac{dx - v \cdot dt}{\sqrt{1 - \frac{v^2}{c^2}}}$$

and by posing $\beta = \frac{v}{c}$, the previous becomes

$$(20) \ dx' = \frac{dx - v \cdot dt}{\sqrt{1 - \beta^2}}$$

In (20) dx' is the length measured by the observer in O' moving solidly with the ruler, and dx is the length of the ruler in motion, measured by the observer in O at rest.

In order for observer O to perform the ruler measurement, it is necessary for him to perform the measurement of the ends contextually (simultaneously), i.e., such that $dt = 0$. Substituting the latter value into (20) we get:

$$(21)\ dx' = \frac{dx}{\sqrt{1 - \beta^2}}$$

and again by placing

$L' = dx'$ (Length of the ruler in O')

$L = dx$ (Length of the ruler in O')

$$\gamma = \frac{1}{\sqrt{1 - \beta^2}}$$

(21) becomes

$$L' = \gamma\, L$$

which rewritten in the inverse form becomes

$$(22)\ L = \frac{L'}{\gamma}$$

From that relationship, $v < c$ always being $v < c$, $\beta < 1$, as a consequence $\gamma > 1$, and again will always result $L < L'$.

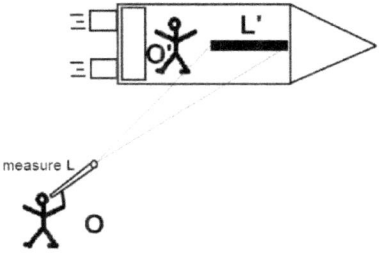

measure L

O

We have derived that the length of an object in relative motion, measured by an observer O at rest always turns out to be less than the length of the same object measured by the observer O' integral with the motion.

It happens that observer O observes a contraction of the lengths of an object in relative motion with respect to it.

2.7 OWN LENGTH

Analogous to proper time, we define Proper length: the length measured in the reference system integral with the observed event or phenomenon, and is always the maximum measurable length of an object.

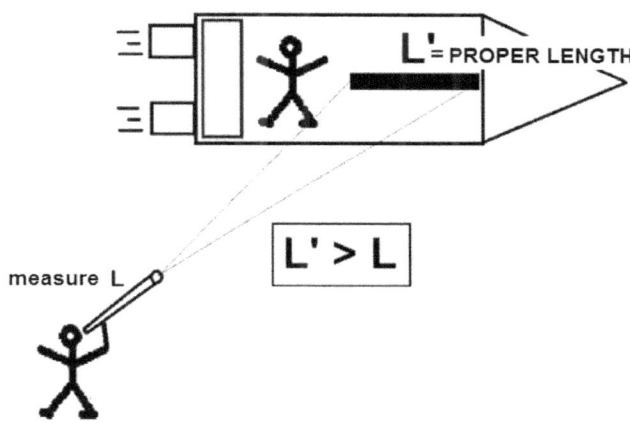

As with time dilation, a mental experiment can also be performed for length contraction. This latter mental experiment is known as the PARADOX OF AUTORIMENT.

Consider a car and a garage, both in quiet, with the garage having a shorter length than the car.

The garage has two doors that can be opened and closed vertically "instantaneously" through a sensor that recognizes when the car inside the garage arrives, with the front, at exit door B and with the rear at entrance door A.

Under the initial conditions of the experiment we place both doors open.

Under such conditions, the car cannot fit inside the garage with the doors closed, due to the initial assumption that the car is longer than the length of the garage.

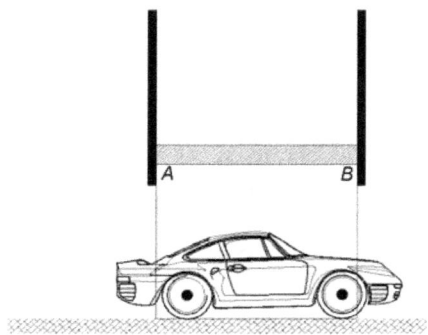

Consider a reference system integral with the garage at rest, then an observer placed in the garage.

By running the car at relativistic speeds (step 1) for length contraction it happens that the car will appear contracted relative to its actual size and thus be contained in the garage (step 2).

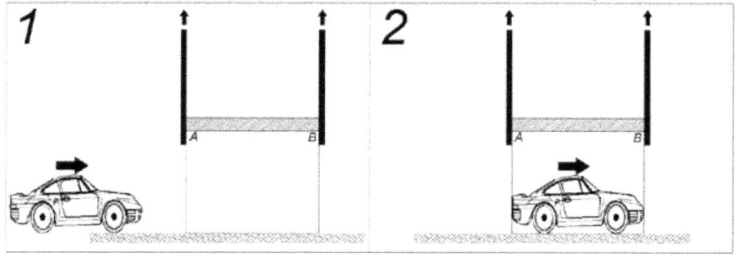

When sensor B detects the front of the car, at the same time sensor A also detects the tail of the car, so the doors close simultaneously (step 3) and an instant later they open simultaneously to let the car continue its run (step 4).

Since both the head and tail of the car were, if only for a moment, simultaneously at the closed entrance and exit doors we can admit that the car is contained in the garage.

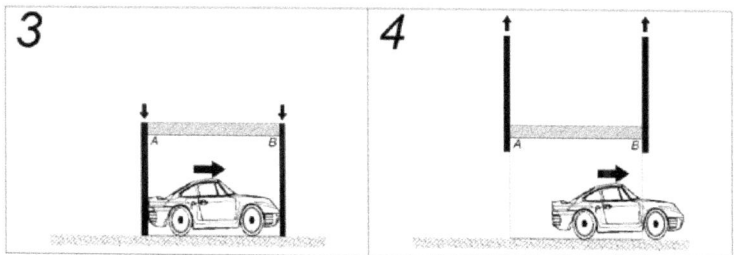

By reciprocity we can think of the same experiment from the car's point of view.

Any phenomenon must be observable from different reference systems without substantial variation in interpretation of the same.

By the principle of reciprocity, we repeat the thought experiment considering the reference system integral with the car.

Under this assumption it will result in the garage moving at relativistic speeds and the car at rest. Consequently, for the contraction of lengths, the garage will result in a length less than its actual value at rest.

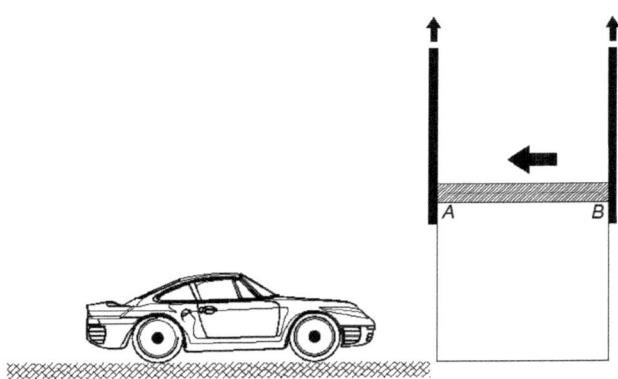

It quickly becomes apparent that the car by metric is not contained in the garage.

This leads to a paradox in that by reciprocity the experiment cannot lead to different results by varying the reference systems for observation.

Let us analyze in detail, however, what happens in this second case, with the garage in motion and the observer quiet in the car.

When sensor B detects the front of the car, door B closes

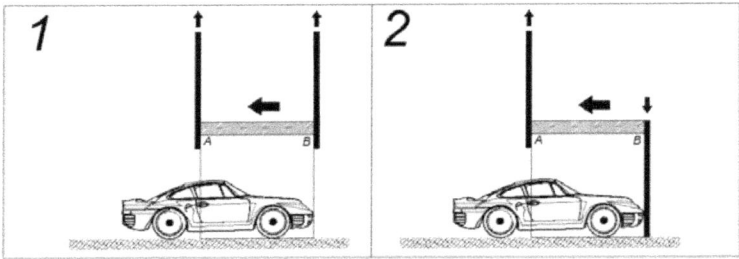

Instantly door B reopens, and when sensor A detects the tail of the car, it closes.

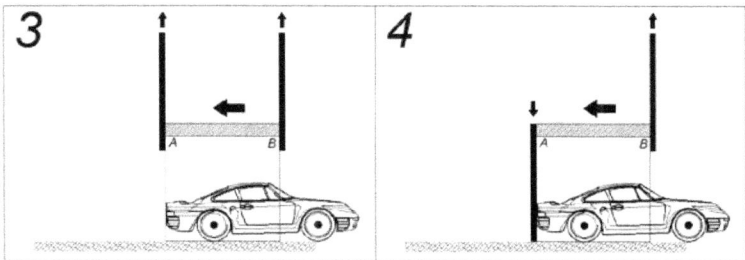

We specify that the doors move together with the garage at relativistic speeds.

The observer, in stillness, placed in the car, due to "time dilation" will measure the occurrence times of the events of opening (B) and closing (A) of the doors in a dilated manner, i.e., he sees time flowing more slowly.

Under such a hypothesis, the time difference between the events of opening (B) and closing (A) of the doors, observed from our reference system integral with the car, occurring in the moving garage are dilated by a factor $\gamma > 1$, such that the events relating to the closing and opening of the two doors appear non-simultaneous.

In fact, the latter is a time that does not exist, as we had observed in the experiment with the quiet garage.

For this reason, the events related to the closing and opening of the two doors appear to be non-simultaneous.

As a result, the car continues to be contained in the garage, since both the front and the rear are at a closed door, although not simultaneously.

Therefore, two same events may be simultaneous or nonsimultaneous in different reference systems, but the physical phenomenon must, without question, have the same unfolding. The concept that two same events can be simultaneous or nonsimultaneous in different reference systems is again reiterated.

2.8 SIMILARITIES WITH OTHER PHENOMENA

Time that expands when a twin departs at great speed on a spacecraft and the lengths of objects that contract when they run are concepts that are difficult to grasp in the rationality of the everyday observable.

Yet, in nature we know and observe, without any perplexity, the thermal expansion of a metal by heating, in the same way that the time of an event in relative motion expands.

An object moving away from us becomes smaller and smaller just as the measure of a length in relative motion is always smaller than its own length.

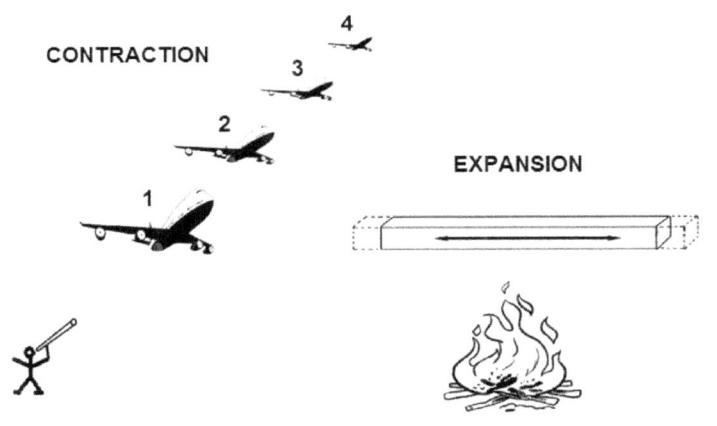

CHAPTER 3
RELATIVISTIC DYNAMICS

3.1 RELATIVISTIC MASS

So far we have dealt with material points and segments without ever having considered the magnitude "mass" of a moving body. We can distinguish two types of masses: rest mass and relativistic mass.

Rest mass represents an invariant and is the mass of a body at rest.

Relativistic mass, unlike mass at rest, which is used without distinction in classical mechanics, can take on multiple aspects in relation to how it is identified.

Relativistic mass, in general, depends on velocity, and as velocity increases it tends toward infinity.

This consequence leads to the admission that a body is impossible from reaching the speed of light, since as the velocity increases, the mass increases more and more, until it assumes such a high value that it opposes any further increase in speed.

A more exact definition of mass in the physical sense follows.

If we want to identify mass as the ratio of force to acceleration, we have a Transverse Mass (m'_T) with the Force placed in the direction transverse to the acceleration.

On the other hand, when the Force is in the same direction as the acceleration we will speak of Longitudinal Mass (m'_L).

It can be shown that the transverse relativistic Mass is equal to:

$$m_T' = \frac{m_o}{\sqrt{1 - \beta^2}}$$

$$m_T' = \gamma \, m_o$$

While the longitudinal relativistic Mass is equal to:

$$m_L' = \frac{m_o}{(\sqrt{1 - \beta^2})^3}$$

$$m_L' = \gamma^3 m_o$$

If, on the other hand, we identify mass as the ratio of momentum to velocity, we obtain relativistic mass more simply.

$$\vec{p} = m' \cdot \vec{v}$$

$$m' = \frac{\vec{p}}{\vec{v}}$$

$\vec{p}$ is referred to as a quadrimpulse as it is a function of the three dimensions space and the fourth dimension time.

Relativistic mass takes the form:

$$(23)\ m' = \frac{m_o}{\sqrt{1 - \beta^2}}$$

$$(24)\ m' = \gamma\, m_o$$

The mass m_o coincides with the mass at rest, that is, the mass of the body invariant for any reference system, since it does not depend on velocity.

For low velocities the rest mass coincides with the mass of the body.

Relativistic mass, on the other hand, is not invariant because it depends on the factor γ and thus on the velocity measured in the reference system under consideration.

From the preceding relation (23) we derive that relativistic mass increases with increasing velocity, allowing us to derive the difficulty of being able to accelerate a particle to the speed of light.

At velocities close to that of light, for example, substituting in (23) a velocity value equal to 99% of the speed of light ($v = 0.99\ c$) the relativistic mass is equal to:

$$m' = \frac{m_o}{\sqrt{1 - 0.99^2}} = 2.525\ m_o$$

The mass of a body in relativistic motion becomes about 2,500 times its original mass, such that, for example, a rest mass grows from 100 kg to 252,500 kg = 252.5 tons.

3.2 RELATIVISTIC ENERGY

In this section we will perform simple calculations to derive Einstein's best-known and simplest relation $E = mc^2$, crappresentative of the mass-energy duality as two sides of the same coin.

This relationship expresses an innovative concept of equating mass and energy. In accordance with the transformation principle, mass can transform into energy and vice versa.

The Kinetic Energy (Energy possessed by a moving body) required to bring a moving body from velocity *v* to a position at rest (stopping the body) is equal to

$$E_c = \int_v^0 F \, dx$$

With $F = \frac{d \, (m \, v)}{dt}$ and $m = \gamma \, m_o$,

By substituting, integrating and performing the appropriate operations, we obtain:

$$(25) E_c = mc^2 - m_o c^2$$
$$(26) E_c = (\gamma - 1) \, m_o c^2$$

Where E_c is the kinetic energy, mc^2 is the total energy E_T and $m_o c^2$ is the energy at rest E_0.

The (25) can also be written

$$(27) \; E_T = E_c + \, m_o c^2$$

Ultimately, we obtained that the total energy E_T is given by the sum of the kinetic energy E_c with the energy of the mass at rest $E_0 = m_o c^2$.

And here appears the famous formula whose appeal lies in its incredible elegance and extreme simplicity.

$$(28)\ E_0 = m_o c^2$$

$$(29)\ E_T = mc^2$$

In particular, the relationship says that the energy of a body is directly proportional to the speed of light squared, such that energy takes on very high values even for small values of mass. For the rest energy we will consider the corresponding rest mass m_o, while for the total energy we will need to consider the relativistic mass.

This ingenious relationship establishes that mass and energy are equivalent.

In the latter relations, fundamental properties of mass (matter) are summarized, which become explicit in the course of interaction with energy:

- mass varies continuously in relation to total energy. Specifically, it increases by an amount equal to E/c^2 when it absorbs energy (electromagnetic radiation), while it decreases when it loses energy, such as by emitting photons; the amount of mass that disappears corresponds to an emitted energy equal to mc^2 (29);

- a body at rest possesses energy merely because it has mass (28);

3.3 KINETIC ENERGY AT LOW SPEEDS

In the case where the velocities involved are low (non-relativistic), the physical laws of classical mechanics continue to apply, even for the calculation of energy.

Analytically, it is obtained that from (26):

$$E_c = (\frac{1}{\sqrt{1-\beta^2}} - 1)\, m_o c^2$$

By serially developing the term with the root

$$\frac{1}{\sqrt{1-\beta^2}} = 1 + \frac{1}{2}\beta^2 + \frac{3}{8}\beta^4 + \ldots..$$

You get

$$E_c = (1 + \frac{1}{2}\beta^2 + \frac{3}{8}\beta^4 + \ldots - 1)\, m_o c^2$$

Considering that the assumed velocities involved are small, it is permissible to neglect the terms following the β^2, as the ratio of the conisdered velocity to the speed of light

$$E_c = \frac{1}{2}\beta^2\, m_o c^2 = \frac{1}{2}\frac{v^2}{c^2}\, m_o c^2 = \frac{1}{2}\, m_o v^2$$

Which assumes the same notation as the already well-known relation for calculating kinetic energy in classical mechanics.

CHAPTER 4

GEODESICS AND MINKOWSKI SPACE-TIME

4.1 SPACE-TIME INVARIANT IN THE CHRONOTOPE

In Galilean systems (reference systems in inertial motion at low velocities), the distance between two points turns out to be an INVARIANT, that is, it does not vary as the reference system under consideration changes.

Distance is defined as the measure of the minimum possible path between two chosen points.

A ruler always measures the same length whether it is in motion or at rest.

In three-dimensional Euclidean space, the distance between two points represents an INVARIANT, and this minimal entity is called a GEODESIC.

When passing into 4-dimensional space, three spatial coordinates in addition to time, the latter not being constant as the reference systems change, what happens is that the distance between two points no longer turns out to be an invariant, that is, it no longer represents a geodesic.

It is therefore necessary, in this circumstance, to introduce a new entity that meets this invariance condition.

In relativistic 4-dimensional space (Space-time) a fourth dimension identified in the time variable is added compared to 3-dimensional space. Such 4D space takes the name PSEUDO-

EUCLIDEAN CHRONOTOPE or as we shall see later Minkowsky's M4 space.

In 4D space, remembering Lorentz transformations, measurements can appear contracted or dilated, depending on the relative state of motion.

The new invariant entity in the PSEUDO-EUCLIDEAN CHRONOTOPE is still a Geodesic, however, representing the shortest path between two points/events instead of between two simple points.

Let us now analyze the above in analytical terms.

We start with Euclidean space, but for simplicity we use a two-dimensional system.

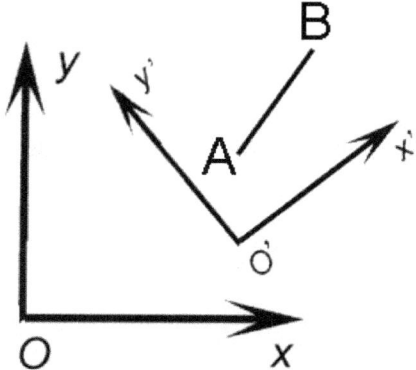

The segment AB will always have the same length regardless of whether the measurement is made from the reference system O or O'.

Its measurement is derived from the well-known formula derived from Pythagoras' theorem

$$\Delta s^2 = \Delta x^2 + \Delta y^2 = \Delta x'^2 + \Delta y'^2 = INVARIANT$$

In three-dimensional space and in differential form:

$$(30)\ ds^2 = dx^2 + dy^2 + dz^2 = dx'^2 + dy'^2 + dz'^2 = INVARIANT$$

Extracting at the root gives

$$(30)\ ds = \sqrt{dx^2 + dy^2 + dz^2}$$

Exactly, the distance between two points represents an invariant entity, defined as Geodesics in Euclidean space.

In 4-dimensional space, Pseudo-euclidean chronotope, (30) is no longer an invariant element.

Another entity must be found that also takes into account the time variable, thus abandoning the concept of position and introducing the concept of EVENT: position in time.

An EVENT point is represented by the three spatial coordinates in addition to the time coordinate.

It is defined as invariant in the Chronotope, the magnitude σ

$$\Delta\sigma^2 = c^2\Delta t^2 - (\Delta x^2 + \Delta y^2 + \Delta z^2)$$
$$= c^2\Delta t'^2 - (\Delta x'^2 + \Delta y'^2 + \Delta z'^2) = invariant$$

In differential form

$$(31)\ d\sigma^2 = c^2 dt^2 - (dx^2 + dy^2 + dz^2)$$
$$(32)\ d\sigma = \sqrt{c^2 dt^2 - (dx^2 + dy^2 + dz^2)}$$

(31) can be written in the form of Einstein's notation of the sum of repeated indices,

$$(32a)\quad |x|^2 = g_{\mu\nu}\, x^\mu x^\nu$$

or

$$(32b)\quad d\sigma^2 = g_{\mu\nu}\, dx^\mu dx^\nu$$

which is equivalent to

$$d\sigma^2 = \sum_{\mu,\nu=1}^{4}\left(g_{\mu\nu}\ dx^\mu dx^\nu\right)$$

$$con \quad g_{\mu\nu} = \begin{cases} 1 & se\ \mu = \nu = 1 \\ -1 & se\ \mu = \nu = 2,3,4 \\ 0 & in\ the\ other\ case \end{cases}$$

and then

$$d\sigma^2 = (dx^1)^2 - (dx^2)^2 - (dx^3)^2 - (dx^4)^2$$

not to be confused with the index and exponent.

It should be pointed out that in many texts it is possible to find indices identifying each coordinate in the form 0,1,2,3 instead of 1,2,3,4, without changing anything in substance.

The magnitude $g_{\mu\nu}$ can be written in matrix form, as follows:

$$g_{\mu\nu} = \begin{bmatrix} 1 & 0 & 0 & 0 \\ 0 & -1 & 0 & 0 \\ 0 & 0 & -1 & 0 \\ 0 & 0 & 0 & -1 \end{bmatrix}$$

and represents the metric of Minkowsky or Pseudo-euclidean flat 4-dimensional space.

Let us not forget that the regime is still one of Restricted Relativity, meaning we are dealing with inertial systems without the presence of gravity, hence the appellation "flat space."

Analogous to the distance between two points, which turns out to be the shortest path, the magnitude $d\sigma$ represents the shortest trajectory in 4-dimensional space and is also referred to as the Geodesic.

Recall that in Euclidean/Galilean space the Geodesic reduces to the distance between two points, reducing to zero the component dt.

The metric of 3-dimensional space, in this case is simplified to:

$$g_{\mu v} = \begin{bmatrix} 1 & 0 & 0 \\ 0 & 1 & 0 \\ 0 & 0 & 1 \end{bmatrix}$$

Consequently, from (32b) we again obtain the formulation in (30):

$$d\sigma^2 = (dx^1)^2 + (dx^2)^2 + (dx^3)^2$$

The relativistic invariant in the Pseudo-euclidean Chronotope defined by (32) has with respect to the invariant in Euclidean space (30) an additional term $c^2 dt^2$.

The latter term is homogeneous, however, in that dimensionally it takes on the appearance of a distance, since it is the result of the product of a velocity times a time.

4.2 GEODESIC

To better understand the concept of geodesics in 4-dimensional space, we can examine some analogies in 3-dimensional space. The earth's surface, a three-dimensional entity, can be represented through appropriate projections on a flat, two-dimensional map; in practice, the world map is "stretched" on a flat surface.

Let us examine the trajectory of an airplane moving from point A to point B.

This path is always the minimum distance to be traveled between two points in order to optimize flight time and fuel expenditure.

However, the aircraft trajectories that are normally observed on geographical maps, thus in two-dimensional, appear curved and seem not to be the shortest possible.

This happens precisely in the transition from a spherical, three-dimensional surface to the flat, two-dimensional representation.

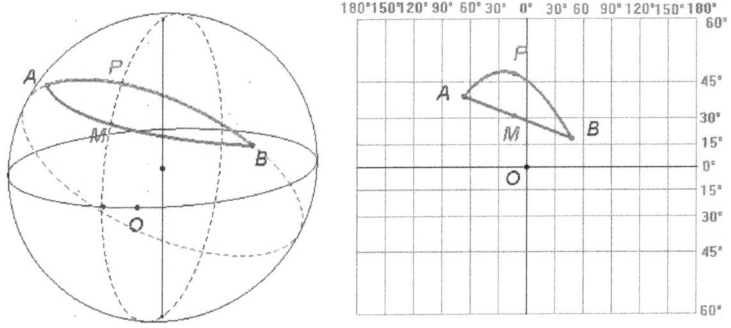

The trajectory P shown in the sphere is the minimum distance between the two points (geodesics) in the three-dimensional real system, i.e., it is a maximum arc of a circle.

The curve P drawn in the plane system, while being the shortest distance in the real three-dimensional system, appears curved to us and much larger than the minimum distance denoted by M. Only when the segment M is returned to the sphere does one gain evidence of the previous distortion in the plane representation, agreeing to move away from conventional visual evaluations that lead to erroneous inferences.

Even in four-dimensional (x,y,z,t) space, the geodesic will be represented by a curve, rather than a line, joining two points-events.

4.3 CALCULATION OF PROPER TIME WITH GEODESICS

Own time is the shortest possible time, as previously defined.
Since geodesics is the minimum distance between two events, it
follows that the proper time will necessarily have to be "the time
of an event traveling a geodesic."
Writing (32) in non-differential form.

$$(33) \; \sigma = \sqrt{c^2 t^2 - (x^2 + y^2 + z^2)}$$

and considering two reference systems in relative motion with
each other, one of which is integral to the event to be observed
(O') and the other at rest (O)

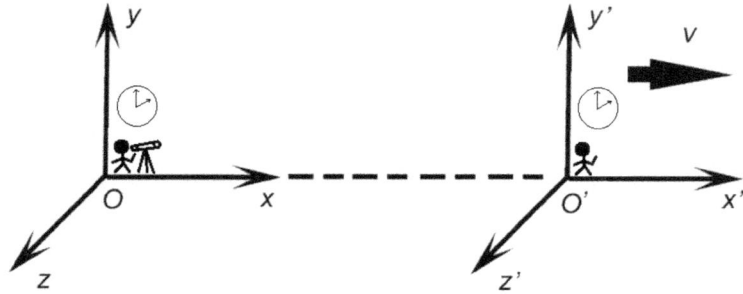

the quantity being σ invariant, we obtain:

$$\sqrt{c^2 t^2 - (x^2 + y^2 + z^2)} = \sqrt{c^2 t'^2 - (x'^2 + y'^2 + z'^2)}$$

Since the reference system O' is integral with the event to be
observed, we have

$$x'^2 + y'^2 + z'^2 = 0$$

and then

$$c^2 t^2 - (x^2 + y^2 + z^2) = c^2 t'^2$$

from which it follows that time t' (proper time = τ) is always less than time t, in perfect agreement with what was stated earlier:

$$\tau < t$$

4.4 MINKOWSKI DIAGRAM

Minkowski space-time is a mathematical model of space-time in four dimensions, of special relativity, in a "local" or "flat" version (without gravity), which can be used to approximate space-time around an event. It is named after its creator Hermann Minkowski, who was among other things Einstein's teacher in Zurich.

The use of such Minkowski space to describe physical systems over infinite distances applies only in the limit of systems without significant gravitation.

In the case of significant gravitation, space-time becomes curved, and one must abandon special relativity for more complete general relativity.

The following discussion is exclusively physical-mathematical.

For simplicity we consider as usual an event in only the spatial dimension (x) as well as the time dimension (t), neglecting the other spatial dimensions (y and z).

Under this assumption, (31) is reduced to.

$$(34)\ d\sigma^2 = c^2 dt^2 - dx^2$$

In non-differential terms

$$(35)\ \sigma^2 = c^2 t^2 - x^2$$

We represent this in a left-handed orthogonal Cartesian axis system showing on the ordinate the velocity value for time ct and in abscissa the spatial value x, examining its physical meaning.

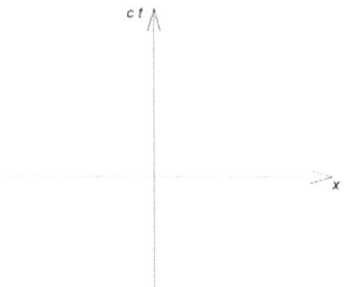

The y-axis *ct* represents a set of stationary events, immobile at position *x=0*.

In fact, by traversing the ordinate *ct,* the spatial position, represented by the abscissa, remains unchanged as time passes.

The same applies to any other vertical line parallel to the *ct* y-axis.

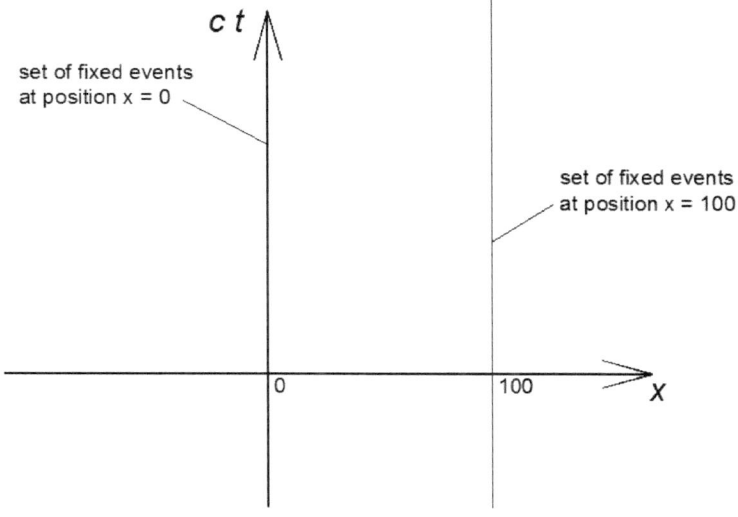

Let us analyze what happens by assuming inclined straight lines. For simplicity we will refer only to the first quadrant with angles ranging from 0° to 45°. The same discussions can be extended to the other quadrants.

A line inclined at 45°, shall, by definition, have the value of the abscissa always equal to the value of the ordinate.

Analytically it is equivalent to equating ct to x in (35), thus obtaining:

$$\sigma^2 = 0$$

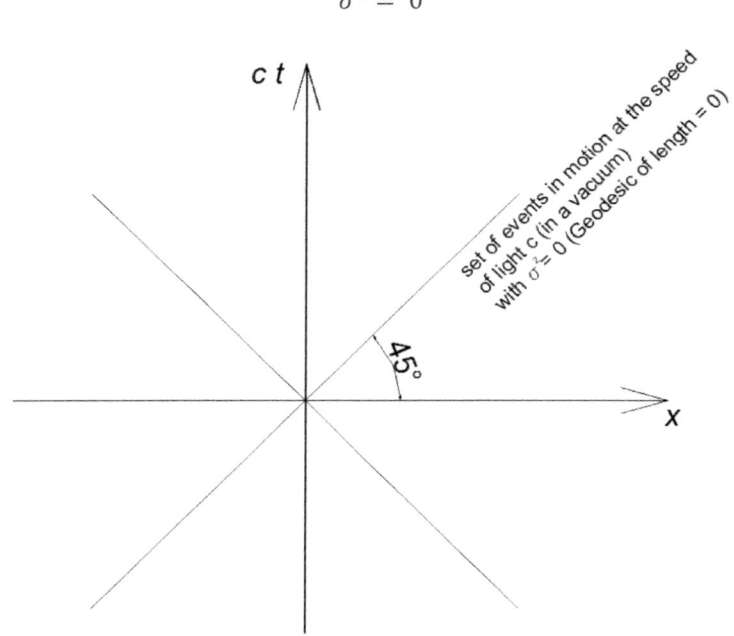

The equality $ct = x$, in the inverse formulation becomes $c = \frac{x}{t}$ indicating, in the kinematic interpretation, a set of events traveling at the speed of light c (in vacuum).

In such a diagram, therefore, straight lines inclined at 45° represent events moving at the speed of light c, and having zero geodesics ($\sigma^2 = 0$).

Inclined straight lines with angles > 45° are characterized by

$$(36)\; c\, t > x$$

(34) becomes

$$\sigma^2 > 0$$

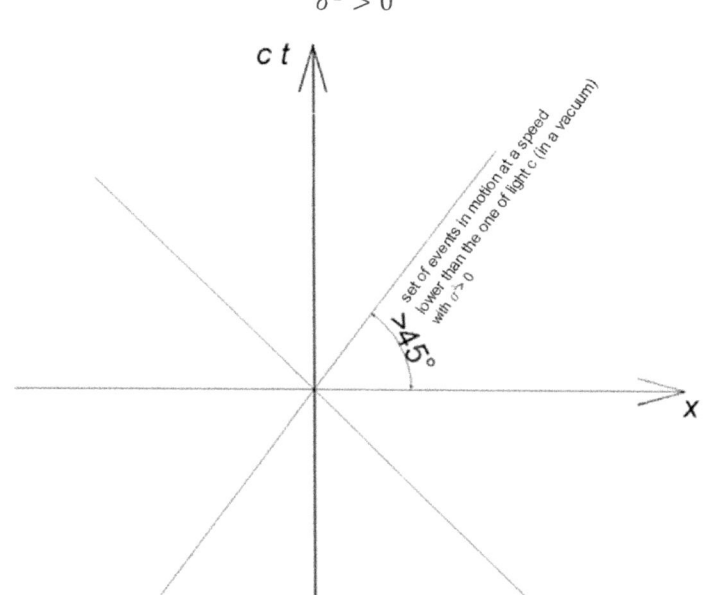

Writing (36) in the reverse formulation. $\frac{x}{t} < c$, we show the representation of the set of events at velocities lower than those of light in vacuum.

Inclined lines of angles greater than 45° represent events moving at the speed less than light c, and having geodesics greater than zero ($\sigma^2 > 0$).

Lastly, we represent the straight lines with slope less than 45°, cartterized by inequality that follows

$$(37)\; c\, t < x$$

(35) becomes

$$\sigma^2 < 0$$

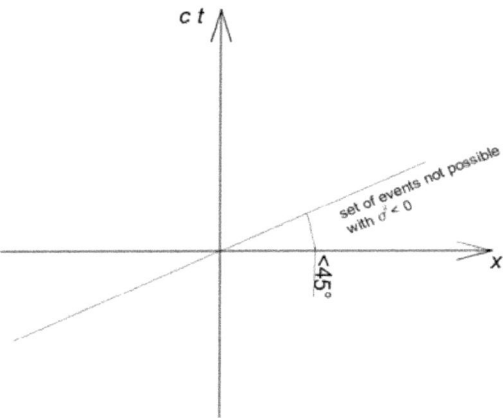

By restating (37) in the reverse formulation. $\frac{x}{t} > c$ it is highlighted from a mathematical point of view how such straight lines represent the set of events at speeds greater than the speed of light in vacuum and therefore not possible.

It is now possible to represent the above spatiotemporal types in a summary graph that distinguishes all possible cases for an event.

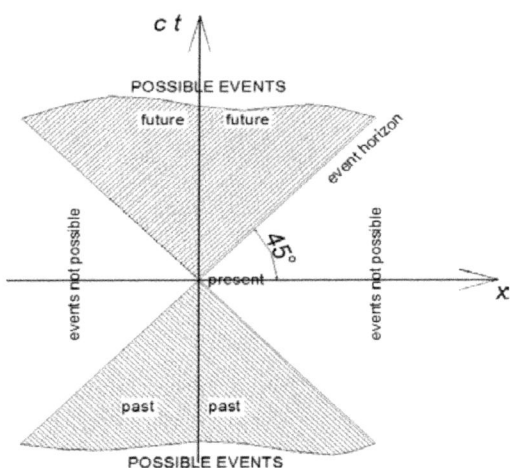

If instead of considering only the spatial coordinate x to graphically represent the set of possible events, we also add the spatial coordinate y, **we obtain a three-dimensional type of graphical representation**

LIGHT CONE

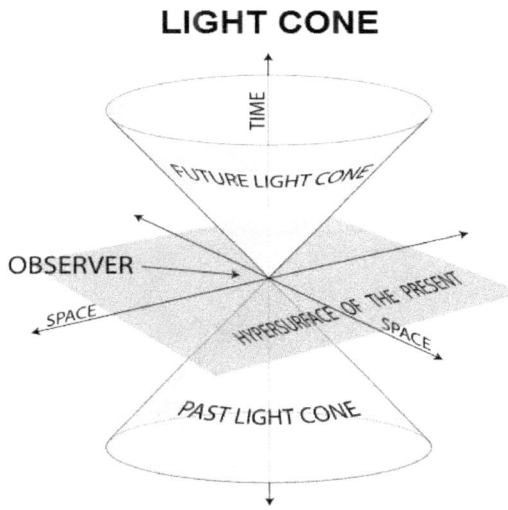

The volume inside the cone represents the possible events (past and future), the horizontal plane represents the present, and finally all the volume outside represents the set of events that are not possible.

If we attempt to complete the representation by introducing the additional spatial coordinate z, things become excessively complicated, as we would have to perform the impossible representation in a 4-dimensional space (x,y,z,t).

4.5 SIMULTANEOUS EVENTS IN THE MINKOWSKI DIAGRAM

We take up the case of "simultaneity in the light event," examined in the previous paragraphs, to carry it out this time with the application of the MINKOWSKI diagram.

Consider the usual train in relative motion with respect to two persons at rest.

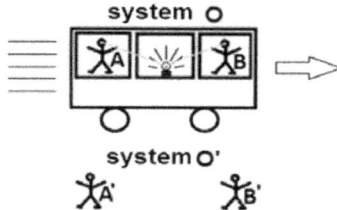

Since the event takes place on the train, by reciprocity, we can consider as the reference system at rest the reference system O, sympathetic to the carriage, and as the reference system in motion O' sympathetic to the observers moving with velocity -v,.

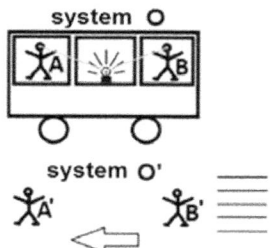

Let us represent this on the **MINKOWSKI** diagram, observing the event from the reference system O, integral with the place of occurrence of the event. Recalling that, for the purpose of Cartesian representation, a light signal is represented by a line inclined at 45°, a body at rest by a vertical line and a body in motion by a line inclined at an angle greater than 45°.

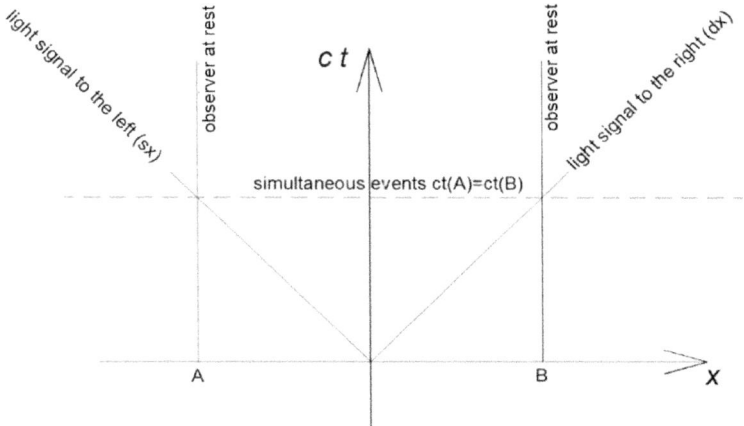

It is easy to read the simultaneity of the event for both observers A and B.

We now evaluate the simultaneity condition of the event for observers A' and B', always observed from the reference system O at rest.

Since A' and B' are in motion in the left direction at a velocity v, they will be represented in the diagram by inclined straight lines with a slope greater than 45° / -45° from the original positions A' and B'

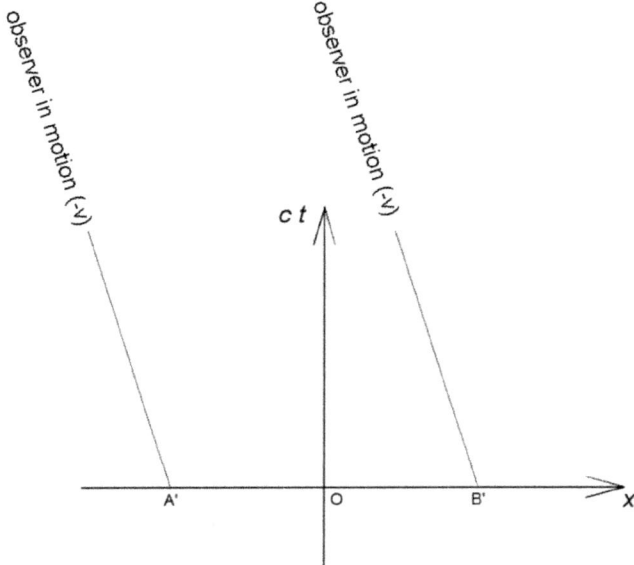

The light signal starting from O will reach A' and B' at different times

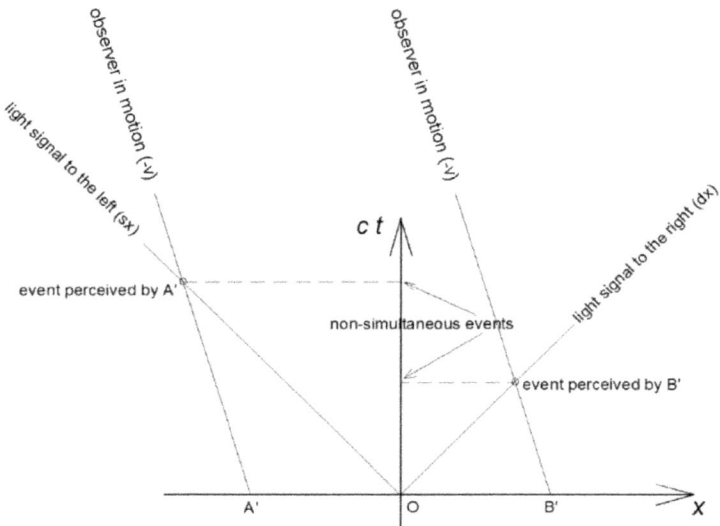

and thus the perception of the event for A' and B' is no longer simultaneous.

Now, however, let us proceed in evaluating the simultaneity condition of the event observed from the reference system O', that is, from the moving system integral to observers A' and B'. We need to consider a new Cartesian axis system having ordinate ct' and abscissa x'.
The ordinate ct', since the reference system O' is integral to A' and B', will have to be inclined by the same angle as the line used to represent the moving observer A' and B'.

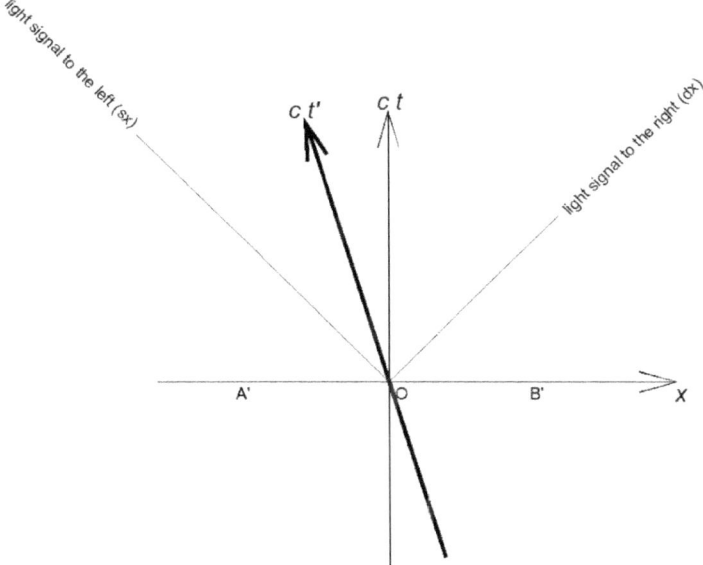

To derive the direction of the abscissae, it is necessary to consider that the light signal will still have to be a bisector line even with respect to the new reference system, the light signal being characterized by having zero geodesics.

The x' axis is derived by plotting the mirror axis to the ct' axis with respect to the light signal line.

Notice how the new Cartesian axes of the moving system are no longer orthogonal to each other.

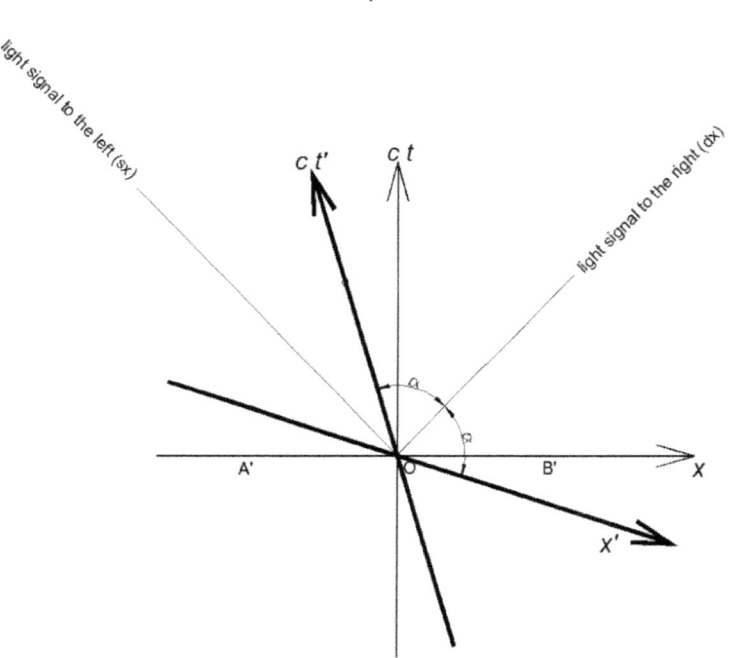

Bringing observers A' and B' (moving) back again.

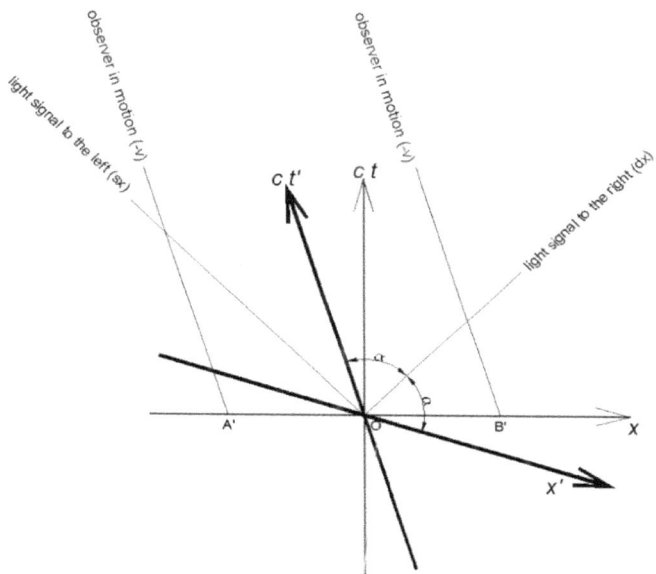

It is obtained that the event observed by A' and B' is simultaneous in the reference system O' with ct'(A') = ct'(B')

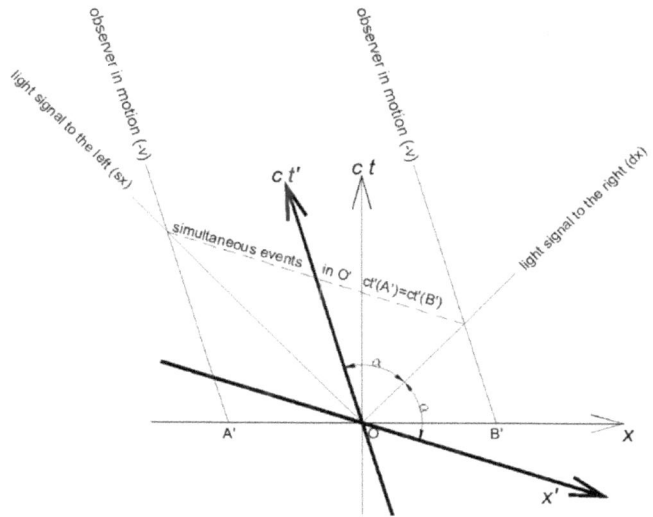

In contrast, the event continues to be nonsimultaneous in the reference system O being, in that reference system, the event perceived at different times t(A') and t(B').

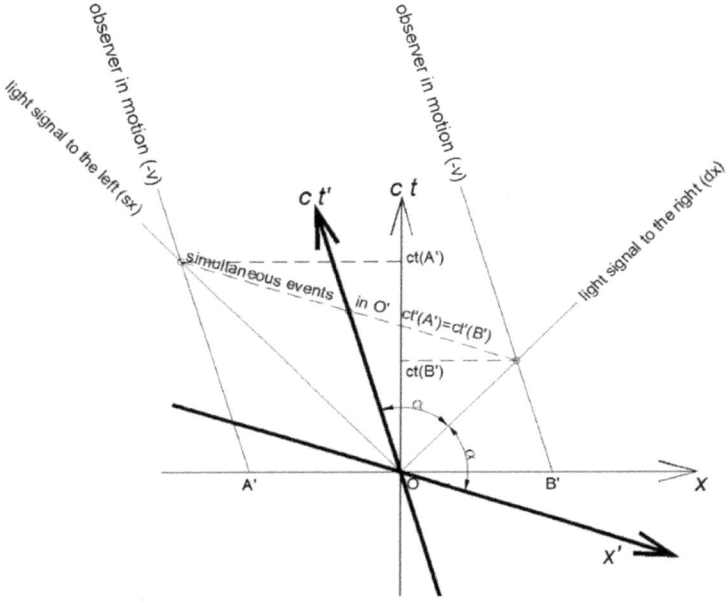

Ultimately, the result was also obtained through the use of the Minkowski diagram that a simultaneous event for two observers placed in one reference system may not be simultaneous for another reference system in relative motion with respect to the previous one.

4.6 TWINS PARADOX ON THE MINKOWSKI DIAGRAM

Let us examine through the use of the Minkowski diagram the usual two observers, with A stationary in the system O and the other B in relative motion integral to the reference system O'

We begin by observing the event represented by B on the spacecraft, from the reference system O at rest with respect to observer A.

At time t=0 both positions of A and B coincide with the origin of the Cartesian system. Next, B moves at velocity v, and we will represent its motion with a line inclined at an angle greater than 45°.

In time $t(A_1)$ the Oobserver B reaches position x_1 .

Meanwhile, the line representing observer A at rest, given that his position does not vary with time, will be precisely the vertical axis ct.

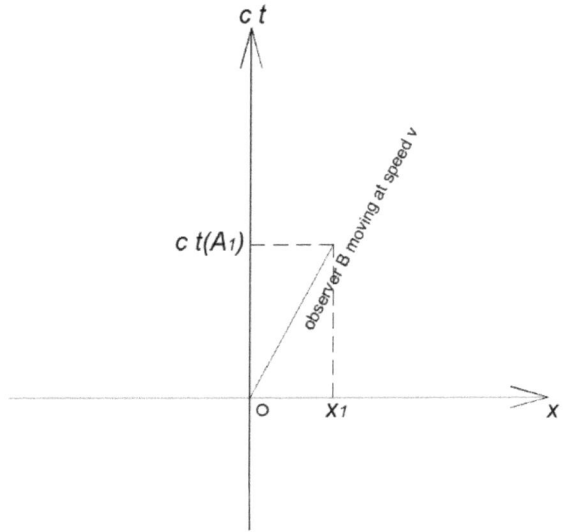

At this point observer B will have to reverse course and return to the original position to meet up with A.

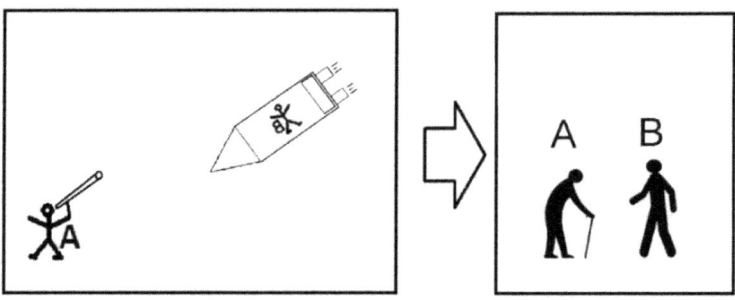

For simplicity, let us consider that an "instantaneous" course change of 360° occurs.

Under such an assumption, the line representing the moving observer B, given the reversal, will bend to the left by an angle opposite to the previous one, until it returns to meet up with A (represented by the vertical axis *ct*) in time t(A$_2$).

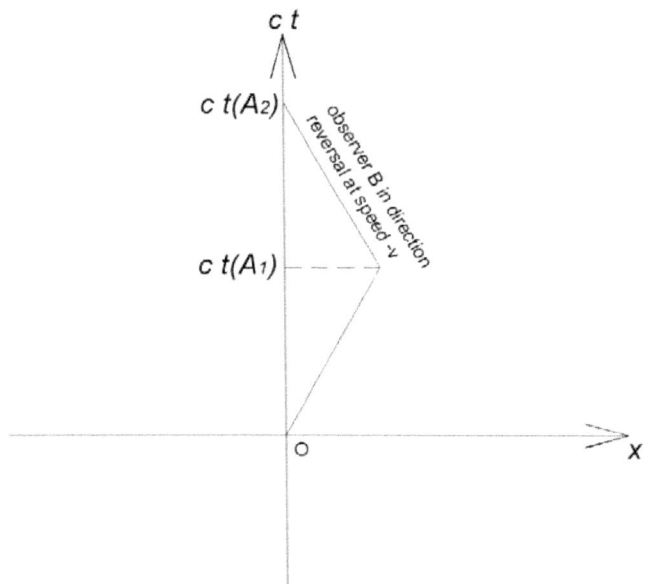

From the reference system at rest O, sympathetic observer A will observe the departure and return of observer B in a time equal to t(A$_2$).

Let us now see what happens by examining the same evemto from the reference system O' integral with traveler B.

On the outward journey the reference system O' integral with B will be represented by an ordinate axis ct' coincident with the line of events at velocity v. The x-axis x' will obligatorily have to be mirrored with respect to the light events, already represented here by a line inclined at 45°.

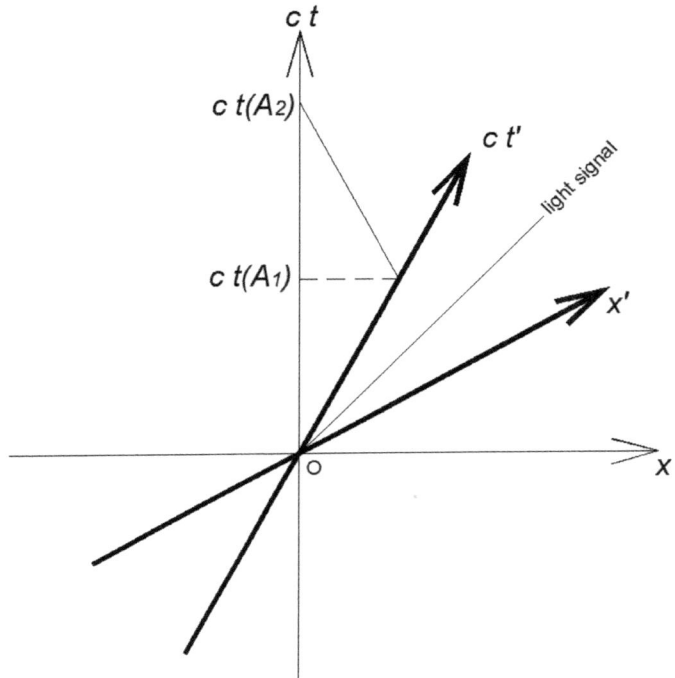

When the traveler reaches point x1 he will have to reverse his course, and as a result the reference system O' integral to B will be rotated in the corresponding direction.

We also assume an instantaneous reversal here.

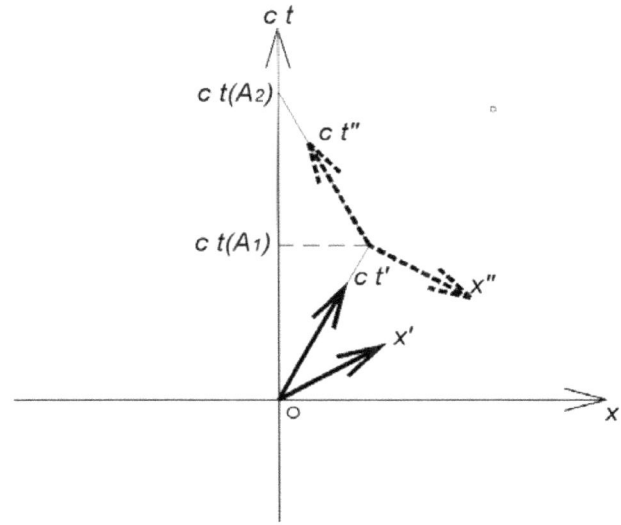

Translating both reference systems to the point of inversion and extending the x-axis to return the times measured by B to the reference system O.

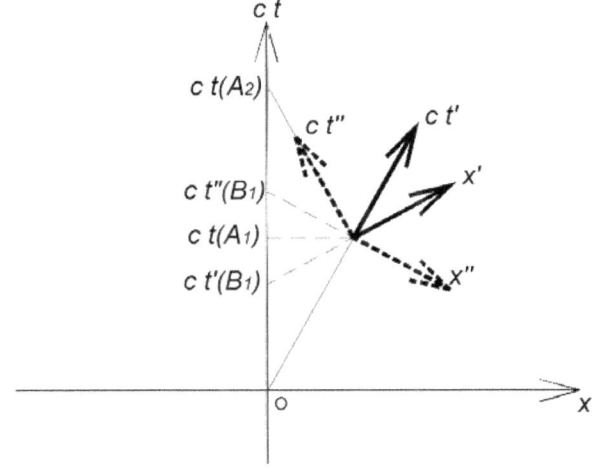

It is obtained that observer B in x_1, a moment before reversing course, measures a time t'(B1) and a moment after reversing course measures a time t"(B1), such that it can be inferred that the time between t'(B1) and t"(B1) does not exist.

Hence its measurement of observer B's time, in motion, will be less than that measured by A precisely by a value equal to the difference t"(B'1) - t'(B1). As a logical consequence on its return B will be younger than A.

All in perfect agreement with what has already been analytically calculated with time dilation, that is, observer A in the reference system at rest measures a time dilated relative to observer B in relative motion with respect to it.

4.7 ALL HERE IS RESTRICTED RELATIVITY

Relativity is represented by everything we have told ourselves in the preceding paragraphs, if we do not take into account the cut-off of a good dose of higher-level mathematics, that is all!!!

We have seen how relativity understood as the study of bodies in relative motion was an issue already addressed by Galileo.

Only later with the advancement of studies on electromagnetism (light) does it happen that using Galileo's physical laws something did not add up.

Galileo's formulas applied to physical phenomena with high velocities involved, comparable to that of light, were no longer valid.

To get down to specifics, what was happening was that Maxwell's formulas (equations) on electromagnetism were no longer valid as reference systems moving at the speed of light changed, that is, Maxwell's equations did not turn out to be "invariant."

Dutch physicist **Hendrik Antoon Lorentz** intervened to resolve the issue, and through an exclusively mathematical artifice he found suitable new relations to restore invariance even in systems related to the speed of light.

Although the Lorentz relations worked perfectly, a physical explanation about them was still lacking.

And this is where a young German physicist, naturalized Swiss and U.S., half-unknown employee at the patent office in Bern, resolved the issue by abandoning the well-established prejudice about the existence of universal time.

Einstein introduces, in addition to the constancy of the speed of light, the existence of different (relative) times based on the motion of the observed system or the observer himself.

An event can be simultaneous with respect to some reference systems and simultaneously nonsimultaneous for other reference systems.

It so happens that Einstein qualifies and quantifies measurements of mysteriously dilated or contracted times and lengths through simple mathematical reasoning and equations, deliberately not reported in the present discussion.

Having conditioned the motion of every possible material body to the maximum value of the speed of light also enabled him to formulate the elegant and famous law $E = mc^2$.

The latter law reveals the mysterious property of equivalence between mass and energy as two sides of the same coin.

The limitation of Narrow Relativity is given by the simplifying condition that the reference systems are in motion, albeit at high velocities, with uniform rectilinear motion, that is, without acceleration and/or curvature.

Another limitation is not to consider too large (massive) bodies, such as stars.

Such reductive conditions would lead Einstein to continue his studies of relativity until he arrived at the wonderful formulation of General Relativity set forth in the following paragraphs.

Ultimately, special or special relativity is not visible in the actions of our daily lives, since we move at speeds at most equal to that of an airplane, and so we will never see our twin returning from a journey younger than we are.

Restricted relativity is actually a very valuable theory, full of mathematics, usable only by those in the field.

Narrow relativity makes it possible to use the satellite navigation system, through the implementation of the GPS system and the possibility of high-precision calculations of luminary signals communicating with satellites placed high in the atmosphere.

Special relativity allows us to make calculations at the level of atomic particles in the course of nuclear reactions. If we wanted to make an atomic bomb we would certainly need to know the theory of relativity.

CHAPTER 5
GENERAL RELATIVITY

5.1 A UNIVERSAL LAW

So far we have been concerned with researching the invariance of physical laws in inertial reference systems without worrying about accelerated systems and again, neglecting gravity.

The theory on General Relativity was born with the aims of seeking a universal law that is valid in every reference system without any limitation.

Einstein worked for more than 10 years on solving what he set out to find.

In 1915 he finally proposed an equation, now known as Einstein's field equation.

Einstein's field equation is a nonlinear partial derivative differential equation and is the foundation for cosmological studies.

5.2 GRAVITY.

Gravity as formulated by Newton consists of a phantom action at a distance that intervenes in the attraction of massive objects.

The force to which it gives rise in the presence of bodies having masses m_1 e m_2 placed at a distance d, is given by:

$$(38)\ F = G\ \frac{m_1 \cdot m_2}{d^2}$$

With G universal gravitational constant having a value of $6.67 \cdot 10^{-11}$ m / Kg s^2 .

The radius of action of the Gravitational Force is infinite, and as can be seen from the above relationship, it is directly proportional to the mass of the bodies and inversely proportional to their distance squared.

It should be noted that the law of Universal Gravitation as formulated by Newton is still valid and used for multiple cosmological predictions, particularly within the limits of our solar system.

So much so that, with the help of that theory in 1846, through appropriate calculations the distant planet Neptune was discovered.

Before proceeding, it is good to realize our size in relation to our surroundings, at least in the immediate cosmological vicinity, through reading the out-of-scale diagram that follows.

113

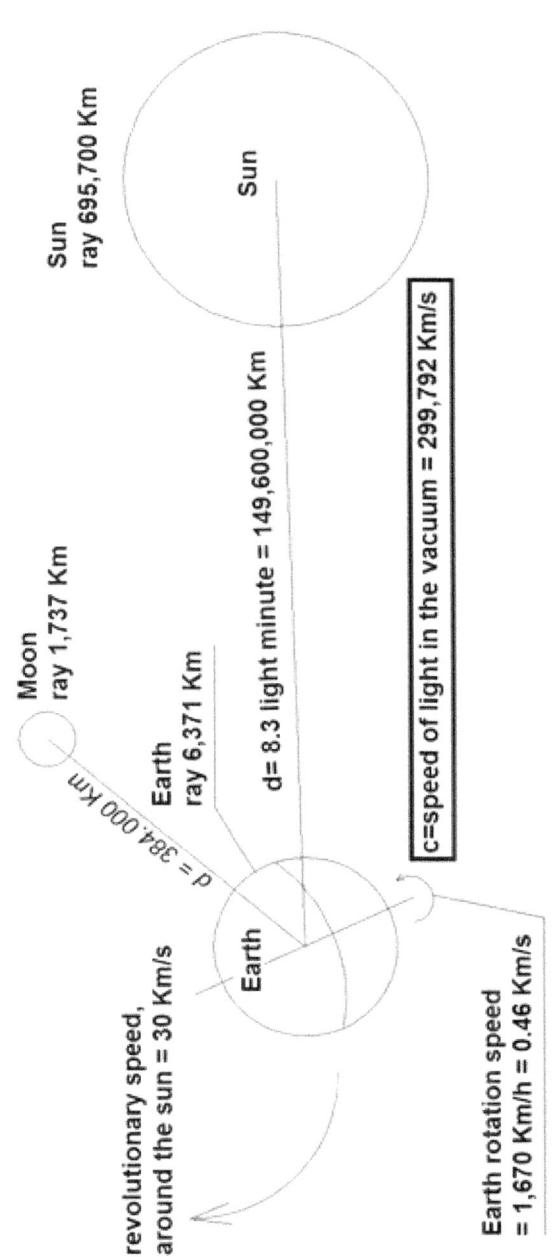

Sun
ray 695,700 Km

Sun

Moon
ray 1,737 Km

Earth
ray 6,371 Km

d= 8.3 light minute = 149,600,000 Km

d = 384,000 Km

Earth

revolutionary speed,
around the sun = 30 Km/s

Earth rotation speed
= 1,670 Km/h = 0.46 Km/s

c=speed of light in the vacuum = 299,792 Km/s

Let's assume that an astronaut wants to undertake interstellar travel to places in weightlessness, although it is clear that it would be somewhat difficult, given the distances involved and the speeds of today's spacecraft.

Why do we see scenes of astronauts floating in space?

Let's understand it better.

In order for a spacecraft to escape the gravitational pull of the Earth's surface, it must reach a theoretical speed of 11.2 km/s, called escape velocity. Once out of the influence zone of the Earth's gravitational field, the speed can also decrease to save fuel.

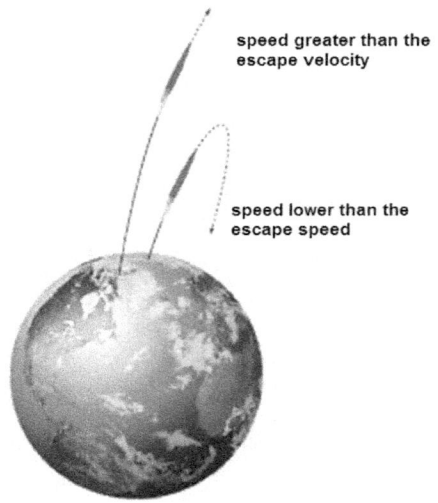

speed greater than the escape velocity

speed lower than the escape speed

To get to the Moon, assuming an average cruising speed of 5 km/s would take about 21 hours.

In reality, the time taken is longer, as cruising speeds are deliberately reduced to decrease fuel consumption and thus refuel less and consequently lighten the ship.

To get to the Sun (assuming we don't melt down the spacecraft) would take about 346 days, refueling permitting.

Astronauts normally reach an altitude on the order of 400 km above the Earth's surface.

Having defined that last altitude, we calculate the difference in the Force of Gravity from the value at the Earth's surface.

Applying (38) in the case of the astronaut and in the case of the earthling

$$(39)\ F_t = G\ \frac{m \cdot M}{r^2}$$

$$(40)\ F_s = G\ \frac{m \cdot M}{(r + 400)^2}$$

With:

$$F_t = gravity\ force\ on\ the\ earth$$

$$F_s = gravity\ force\ at\ the\ spatial\ quote$$

$$m = mass\ of\ the\ astronaut\ or\ man\ on\ the\ ground$$

$$M = mass\ of\ the\ ground$$

$$r = average\ terrestrial\ radius = 6371\ Km$$

Running the relationship between F_s and F_t gives

$$\frac{F_s}{F_t} = \frac{6371^2}{(6371 + 400)^2} = 0,88 \qquad F_s = 88\%\ F_t$$

Ultimately, it turns out that the force of gravity to which an astronaut is subjected is only 12 percent less than the force he was subjected to on the Earth's surface. So he turns out to be just a little thin without any possibility of rising in the spacecraft.

The reason the astronauts are still seen floating in the apparent vacuum is because of the balancing of the force of gravity with the centrifugal force they are subjected to as a result of their rotation around the earth.

Let us now examine the effects of gravity on orbits, applying Newtonian gravitation.

Think of throwing a body from the earth according to Newton's thought experiment.

We know that the (geodesic) trajectory does not depend on the mass from the body, while it definitely depends on the launching speed of the body.

Said body could re-enter Earth's orbit in circular motion or it could continue straight following a small curvature.

It all depends on the launch speed.

We launch a cannonball from the top of a mountain.

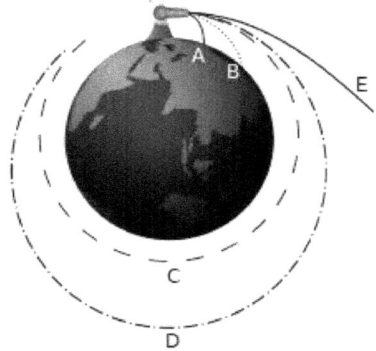

Possible cases can be identified in three categories:

1. The cannonball has such a velocity (and thus a centripetal acceleration $a=V^2/R$) as to result in a "centripetal force" less than the "force of gravity." As a result, the ball falls to earth (cases A and B).

2. The cannonball has such a velocity (and therefore a centripetal acceleration $a=V^2/R$) that it has a "centripetal force" equal to the "force of gravity." The ball enters earth orbit (case C)

3. The cannonball has such a velocity (and thus a centripetal acceleration $a=V^2/R$) as to result in a "centripetal force" greater than the "force of gravity." The ball moves away from the earth (case E)

In the case where the launched body re-enters orbit, that is, proceeds in rotation around the earth, it may happen to assume a motion that is synchronous to the earth's rotation, slower or finally faster.

The above casuistry always depends on the angular velocity (or tangential velocity v= ω r) with which the body enters orbit.

It is possible to calculate the distance of the synchronous orbit, that is, where the body moves of the same angular velocity as the Earth's surface.

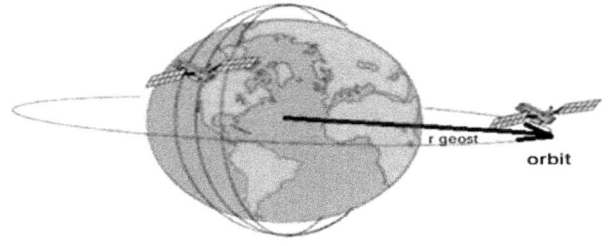

Equating the calculation relation of Centripetal Force (mv^2/r) with the calculation relation of Gravitational Force (38), the desired result is obtained.

$$\frac{m \cdot v^2}{r} = G\,\frac{m \cdot M}{r^2}$$

Simplifying the mass m and introducing the angular velocity ($\omega = \in / r$)

$$\frac{\omega^2 r^2}{r} = G\,\frac{M}{r^2}$$

Simplifying and isolating r to the first member

$$(40a)\ r^3 = G\,\frac{M}{\omega^2}$$

You get

$$r = \sqrt[3]{G\,\frac{M}{\omega^2}}$$

Note how the result is independent of the mass of the moving body, remaining a function of the mass of the planet producing gravity.

Substituting the values of G and M, which for the planet Earth are worth.

G = $6.67 \cdot 10^{-11}$ m / Kg s^2

M = 5.972×10^{24} kg

For the synchronicity condition we assign a value to the angular velocity equal to the earth's angular velocity, that is, we assume that the body makes the same revolutions in the unit time as the planet earth

$$v = \frac{2\,\pi\,r}{T}$$

$$\omega = \frac{2\pi}{T}$$

With

T = Earth's rotation period = 23 h 56 m 4 s = 86,164.00 s

You get

$$\omega = 7{,}29 \cdot 10^{-5} \, rad/s$$

Ultimately, substituting the value of angular velocity thus calculated, we obtain

$$r = \sqrt[3]{G\,\frac{M}{\omega^2}} = 4{,}22 \cdot 10^{-5} m = 42.169{,}00 \, Km$$

Such an orbit is also called a geostationary orbit and is about 6.6 times the Earth's radius, and thus is placed about 36,000 km above the Earth's surface

Below that distance the body rotates faster than the earth, above that distance, on the other hand, the body rotates more slowly, due to the inverse proportionality of the angular velocity with the distance of the rotating body given by (40a).

5.3 INERT MASS AND HEAVY MASS

The foundation of the theory of general relativity is based on the equivalence of inertial mass and gravitational mass, known by the better-known term "weak equivalence principle."

Inertial mass is proportional to the inertia of a body, which is the resistance to change in the state of motion when a force is applied.

To understand it is the mass we find in Newton's law

$$F = m \cdot a$$

Gravitational mass, on the other hand, is proportional to the force of interaction of a body with the gravitational force.

And it is the mass we find in Newton's Law of Universal Gravitation

$$F = G \, \frac{m_1 \cdot m_2}{d^2}$$

Let us perform the following thought experiment in order to clarify the equivalence of the two masses.

Imagine a transparent box containing a guest observer together with other objects, all suspended in empty space (without the presence of gravitational interactions).

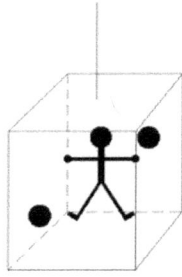

At this point a "prankster demon" pulls the box upward with uniformly accelerated motion (constant acceleration), and all objects, including the observer, having inert mass, placed inside the box begin to move against the bottom of the box.

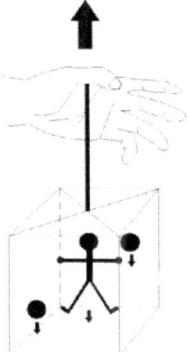

Now imagine the same box placed in the vicinity of a planet and thus suspended in gravitational space.

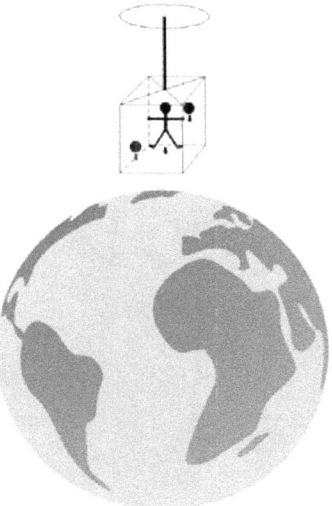

Due to gravitational attraction, all objects, including the observer, having heavy mass will be attracted toward the bottom of the box.

As a result we get that the two effects are completely identical. The two unsuspecting observers placed inside the box cannot distinguish whether accelerated by the prankster demon or by the presence of the massive celestial body.

As a logical consequence we get that the inertial mass can only be equal to the gravitational mass

5.4 TRAJECTORY CURVATURE

Let's continue with some more mental experiments.

We place our box in the vicinity of a planet and using a gun fire a shot.

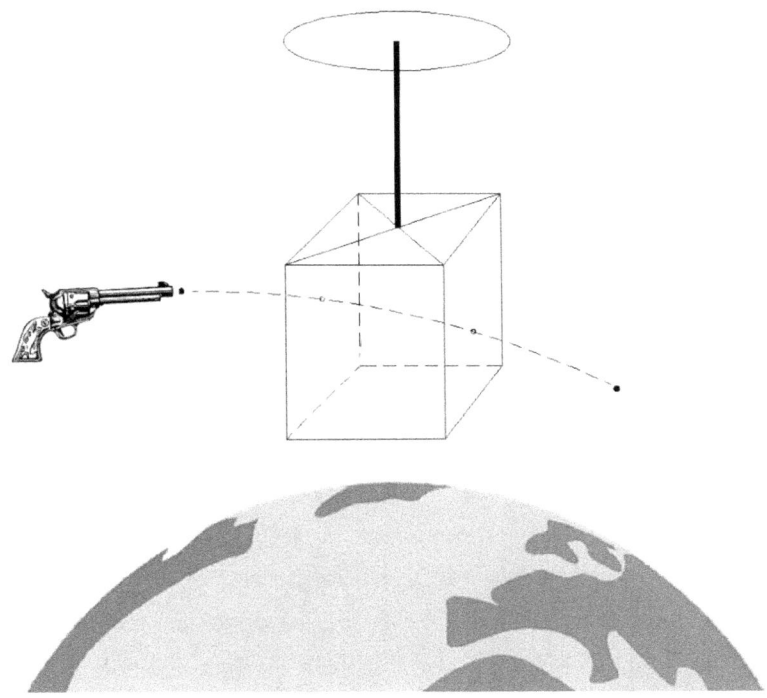

Due to the effect of gravitation, the trajectory of the projectile will be curved as it is attracted by gravitation from the planet.

Now we perform the similar experiment by placing the box in a vacuum, pulled upward by the infamous prankster demon.

In this case the trajectory of the projectile does not curve as it is not attracted by any gravitational field.

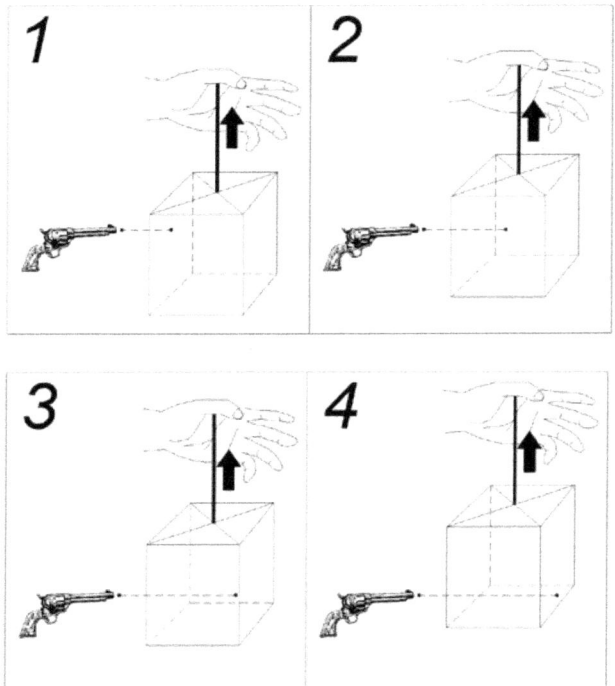

When the projectile punctures the box at the left side, the box being accelerated, until it reaches the opposite side the box will be in a higher position and then it will perform the hole in a lower position.

Instead, an observer sympathetic to the box will still observe a curved-type trajectory.

To visualize said trajectory, simply join the consecutive positions of the projectile following the movement of the box.

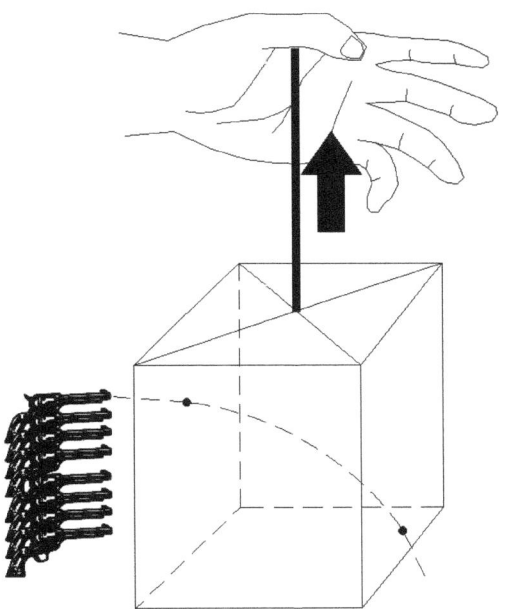

Also in these two cases the equivalence of the curvature of the trajectory is evident and is directly proportional to the acceleration in the case of the prankster demon and by analogy directly proportional to the acceleration of gravity and thus the mass of the planet in the other case.

5.5 CURVATURES OF LIGHT

Performing the latter experiment by replacing the projectile with Light should produce the same result of curvature of the light beam. And that is exactly what happens with the variation that in order to observe the curvature of light, it is necessary to have accelerations of the box such as to produce velocities of the order of magnitude of the speed of light.

By analogy, even in the case where a beam of light is placed in the vicinity of a very massive planet it undergoes curvature.

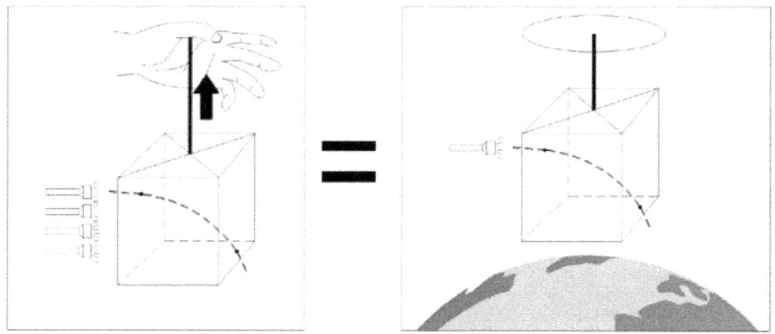

Again, in order to observe the phenomenon of curvature, it is necessary for the gravitational acceleration to be high and thus for the body generating it to be very massive.

The curvature of light in the vicinity of the earth is only a few millimeters, and as a result the bending effects of the light beam are hardly detectable.

The curvature effect can be observed when light interferes with celestial bodies at least as large as the sun, as will be better analyzed analytically below.

5.6 DEFORMATION OF SPACE

What is the light curved by?

By an action at a distance called Gravity Force?

According to the Theory of Special Relativity, the maximum attainable speed is that of light, and therefore there can be no phantom instantaneous force at a distance.

The astounding result Einstein arrives at is that light is not bent by any instantaneous action at a distance, but mass as well as energy deforms the surrounding space. Every particle placed nearby moves in deformed space.

Let us try as always to give examples and analogies.

For it to be possible to proceed with the assistance of graphs, it is necessary to think in a space of only two dimensions.

We think of space as a soft sheet of paper, and mass-energy as marbles of different sizes.

If we place marbles on the sheet (like massive bodies in space), the sheet will deform in proportion to the size of the marble, just as happens with space in the vicinity of massive bodies.

Space deforms by changing the trajectory of every particle, including light, in the vicinity.

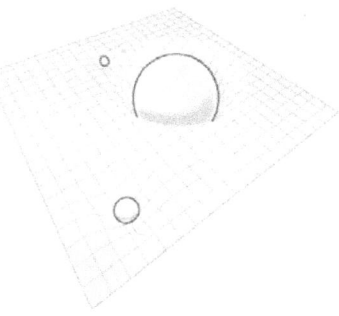

The trajectory of a body or light is still represented by a Geodesic (shortest path), even if a geodesic curves due to deformed space. Such a trajectory always remains the same regardless of the mass of the body.

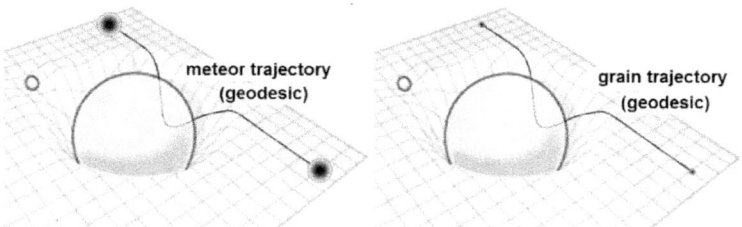

A meteorite approaching a massive body travels the same trajectory as a grain of sand, regardless of their mass in perfect accordance with what Galileo and Newton expressed in the past through their studies of falling bodies and experiments on inclined planes.

Matter tells space how to bend, and space tells matter how to move.

Ultimately, Gravity is only an apparent force, while it is space that deforming itself makes bodies move along geodesics.

The "time" dimension was also deliberately neglected in the previous exposition, which would unfortunately further complicate any graphical representation since it too is subject to deformation.

To better understand the concept, let us continue with more examples.

A body, at rest, placed in absolute space, without influence of other bodies, remains still.

If a body is placed in the vicinity of a massive body, due to space-time deformation, it slides along a geodesic toward this massive body, as if apparently attracted by it.

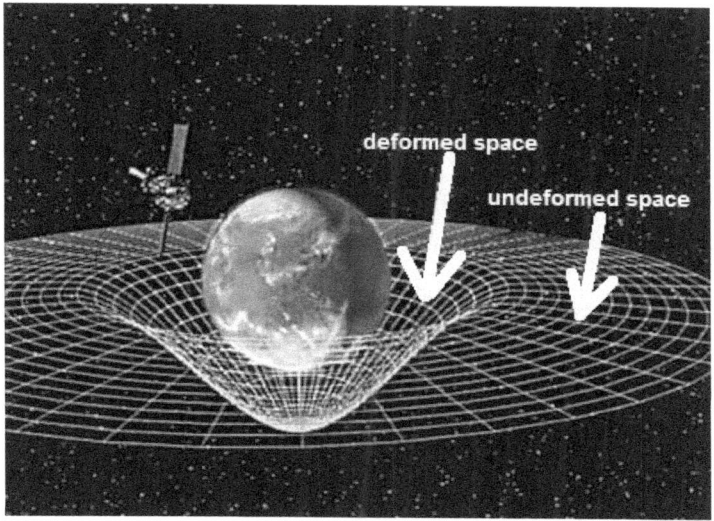

If a body is placed outside the deformed zone, it continues to remain at rest, as it is placed in an undeformed plane space.

5.7 TIME DILATION BY GRAVITATION

Time, like the already seen space, at massive bodies deforms i.e., dilates, in analogy with what happens to time in special relativity

Consider the usual twins.

Twin A, stationary in empty hyperstellar space, far from any massive influence, does not undergo deformation.

Twin B in the vicinity of a massive body, undergoes an influence of space-time deformation.

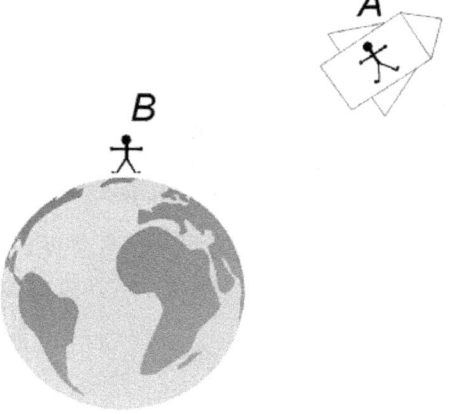

Observer A perceives B's time dilated, as B turns out to be close to a massive body, and by the equivalence principle (massive acceleration = gravitational acceleration) twin B is as if it were moving at increasing speed in an accelerated motion.

With a reference system in A, we have

$$t_B < t_A$$

The time of B being less than that of A turns out to flow more slowly, that is, it turns out to be dilated.

What happens is that if B, after abandoning the twin on the spaceship, approaches the massive body, when he returns and reaches A again, even slowly, he finds him older because A has observed B's time dilated.

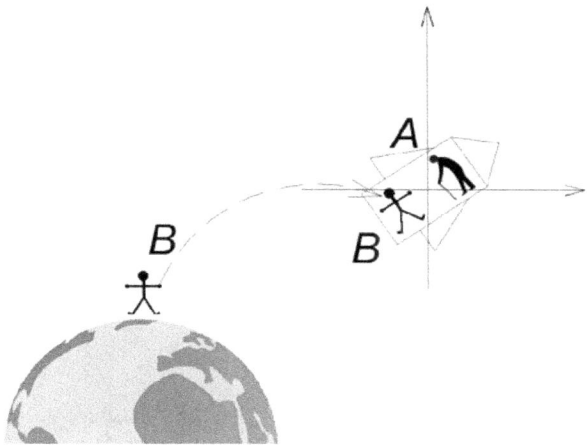

An experimental verification of the above can be performed simply with two clocks, one placed on a skyscraper and one placed on the ground.

In such a condition by difference of the influence of the massive body and thus the gravitational effect, time dilation already becomes measurable. It is obtained that the time of the observer placed on the ground, measured by the observer placed on the skyscraper, is dilated with respect to the time measured by the observer placed on the ground.

The observer on the ground is once again the youngest.

5.8 ANALOGY ON THE DEFORMATION OF SPACE-TIME

To better understand the deformation of space-time we continue with an analogy in a space reduced to the two-dimensional.

Suppose that flat living things, thus two-dimensional, live on the surface of a sphere.

By their very nature, having been conceived "flat," these living beings move only on the surface of the sphere and do not know the third dimension.

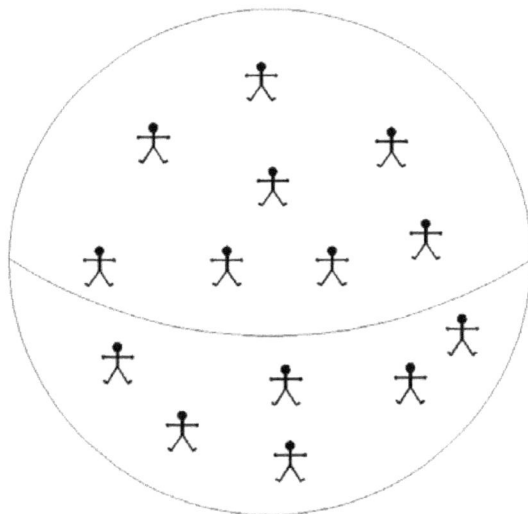

One fine day they decide to build a triangular-shaped dwelling. Therefore, appropriate measurements need to be taken to locate the triangle, as per the plan.

The construction manager engineer proceeds to measure the sides and angles, also verifying, as geometry dictates, that the sum of the internal angles is 180°.

They admit, therefore, that in their space Euclidean geometry is valid and applies.

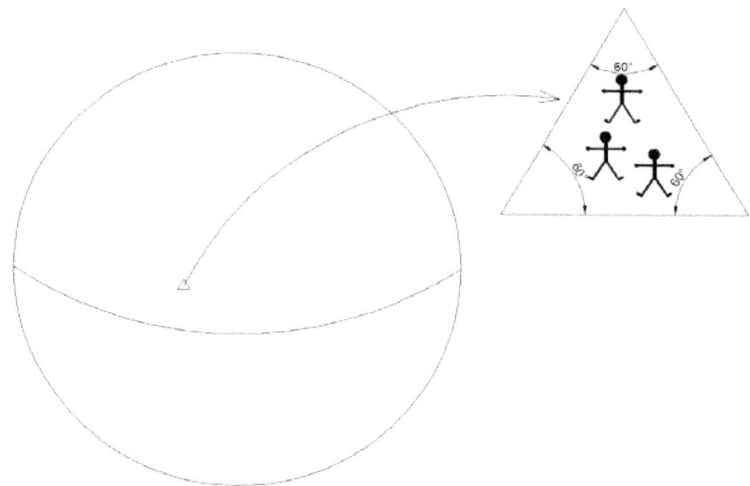

One fine day, one of them named Albert, intrigued, decided to construct and measure a triangle of larger size.

As he proceeds in his measurements, he notices that the sum of the internal angles is greater than 180°, against all geometric predictions.

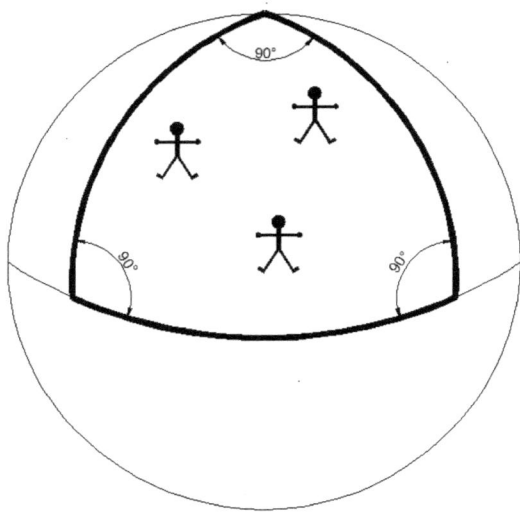

He asserts that Euclidean geometry no longer applies to large triangles.

Indeed, the geometry he finds is distorted into an additional dimension that he does not know.

At this point, in order to justify the unexpected result, all that is left is to invent an apparent force, called Gravipactin, as the cause of the distortion and deformation of known space.

In analogy to the perception in the previous example, General Relativity defines a deformed space in the four space-time dimensions, which we are unable to see on a daily basis because we are limited in our interaction with low-mass bodies and movements characterized by low velocities.

5.9 INTERPLANETARY TRAVEL

Suppose we make an approximate interplanetary section in the immediate vicinity of the solar system.

The following section is schematic and amplified in order to dimensionally enhance the spatiotemporal deformation effects due to the presence of massive bodies.

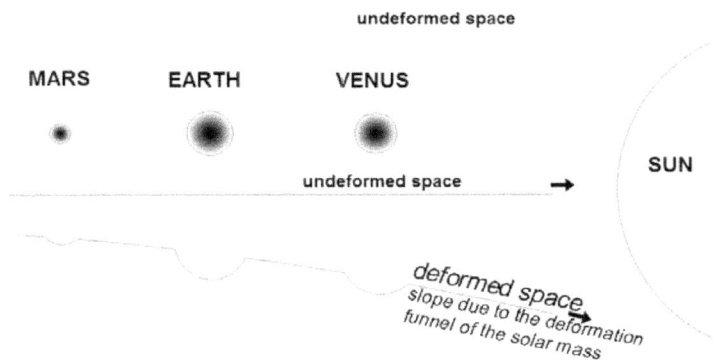

It can be seen from the above representation how the Sun due to its high mass causes a spatial deformation, represented in this case by a line sloping toward the Sun itself.

Each planet also causes a deformation proportional to the respective values of its masses.

A third body that would go into such deformed space, depending on its own motion conditions. will have the option of either sliding toward the Sun or flowing into the deformed basin of the planet.

5.10 AMOUNT OF MATTER IN THE UNIVERSE

Planet Earth appears to have an unlimited surface area because it has no boundaries. Starting from any point in an arbitrary direction, it is possible to move infinitely, even passing several times from the same point.

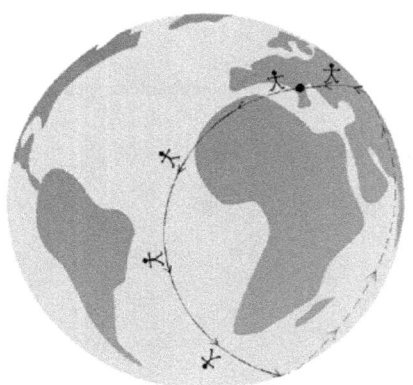

However, the same surface is at the same time "finite." Through a simple geometric formula it is possible to calculate approximately the extnsion of its surface

$$S = 4\pi R^2 = 509,805,891 \ Km^2 = \ 5.10 \cdot 10^8 \ Km^2 \ .$$

Similarly, the volume is also "finite" with value equal to

$$V = \frac{4}{3}\pi R^3 = 1,082,657,777,102 \ Km^3 = 1.08 \cdot 10^8 \ Km^3.$$

By analogy to the previous considerations, the Universe also has an unlimited volume, but at the same time finite and calculable known hypersphere radius. The computational difficulties are greater than in the simpler previous example of the Earth, as calculations in 4-dimensional space (space-time) will be necessary due to the presence of countless massive celestial bodies.

The radius of the hypersphere is estimated to be about 10 billion light-years, or in scientific notation $1.00 \cdot 10^{10}$ al, equal to $9.46 \cdot 10^{22}$ Km .

Immediate consequence of the finite value of the volume of the Universe is that all matter in the Universe is also FINITE.

The previous value of the radius of the hypersphere is calculated as a first approximation from the age of the universe itself, understanding age as the time between the Big Bang and the present.

The age of the universe today is estimated to be 13.82 billion years, equivalent to $4.36 \cdot 10^{17}$ seconds.

Let us analyze a photon that starts in empty space, radially and traces the radius of the hypersphere of the universe.

Such a photon will move along a geodesic in flat space-time, respecting the

$$d\sigma^2 = g_{\mu\nu} \, dx^\mu dx^\nu$$

That in Minkowsky's M4 space and with the origin coincident with the big Bang, becomes

$$\sigma^2 = (c\,t)^2 - (x^2 + y^2 + z^2)$$

Given that the photon is characterized by the null geodesic $\sigma^2 = 0$, and posed for the hypersphere $x^2 + y^2 + z^2 = r^2$, we have

$$(c\,t)^2 = (r^2)$$
$$r^2 = (c\,t)^2$$
$$r = c\,t = 13.1 \ 10^{25} \ m = 13.1 \ 10^{22} Km$$

This value is slightly overestimated because the contribution as a result of time dilation and space deformation was neglected in the simplified calculation due to the imposing presence of mass and energy.

CHAPTER 6
EINSTEIN'S FIELD EQUATIONS
6.1 METRICS OF CURVED SPACE-TIME

Let us return to the thought experiment given in the section on the equivalence between heavy mass and inert mass, and point out that for this equality to be met, it is necessary for the box to be small in size.

In fact, in the hypothesis of a large box within the box, different values of gravity attraction would occur due to the difference in distance from the gravitational source (occurrence of the so-called tidal forces).

Because of this limitation, said equivalence principle is called "weak." In general lindea it is possible to speak of a homogeneous gravitational field only locally.

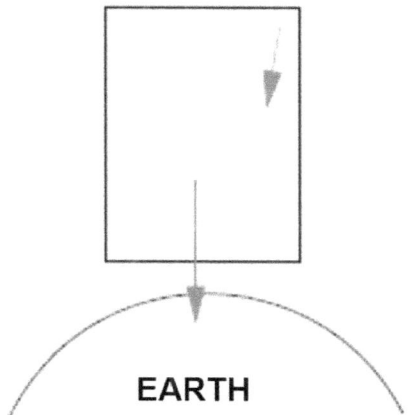

EARTH

Inert mass-heavy mass equivalence, while empirical in nature, was verified by Eotvos and then Dicke with excellent accuracy on the order of 10^{-12} .

It follows from the inert mass-heavy mass equivalence that it is always possible to replace an accelerated system with an equivalent inertial system immersed in a gravitational field.

In any gravitational field, it is always possible to choose a reference system, around a point, where the effects of acceleration due to the gravitational field are zero. This possibility is called the universal or strong equivalence principle.

Ultimately, any accelerated system can be replaced locally by a local inertial system immersed in a gravitational field, due to the weak equivalence principle.

The space affected by a gravitational field, will thus be shattered and replaced by a set of local inertial reference systems, by the strong equivalence principle.

Important considerations flow from the second of the above principles:

- Around a point-event there always exists an inertial map in which the Minkowsky form (flat or pseudo-Euclidean space) is valid

- Space-time with the presence of gravitation, will be covered by many small inertial systems, bound together, however, in a nonlinear way.

In an analogy in 2-dimensional space, we can consider the representation of a curve that can be drawn continuously or subdivided into many segments, neglecting the local curvature of the individual segments while still constituting the segments as a whole a curve. Each segment, however, will be related to the next and the previous one according to a function that allows us to describe the equivalent curve.

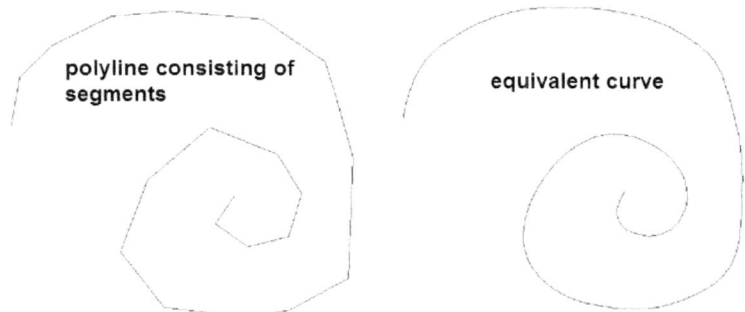

Now applying the strong equivalence principle to two infinitesimal spaces named A and B, which partially intersect, each of which is a locally inertial system.

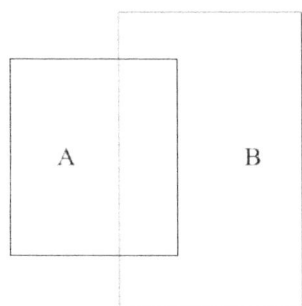

Since A is an inertial system, *(32b)* applies, which written in differential form becomes

$$(40b) \quad d\sigma_A^2 = g_{\mu\nu} \, dx_A^\mu \, dx_A^\nu$$

Since we are considering locally inertial systems, we are in the conditions of pseudo-Euclidean space or Minkowski's metric, and thus

$$g_{\mu\nu} = \begin{cases} 1 & se \; \mu = \nu = 1 \\ -1 & se \; \mu = \nu = 2,3,4 \\ 0 & in \; the \; other \; cases \end{cases}$$

A variety that satisfies a metric describing the distance between two neighboring points of the type *(40b)* with $g_{\mu\nu}$ symmetric and invertible, is called a Riemannian Metric, which is the starting point for the constitution of the structure of space-time that also takes into account the effects of gravity.

In Euclidean space there is biunivocal correspondence between coordinates and points, whereas in a Riemannian variety, as the space becomes curved, the correspondence is only univocal, that is, to a coordinate corresponds a point-event, to a point-event can correspond multiple coordinates.

The behavior is analogous to the metric of a sphere with a center in the origin of a polar coordinate system.

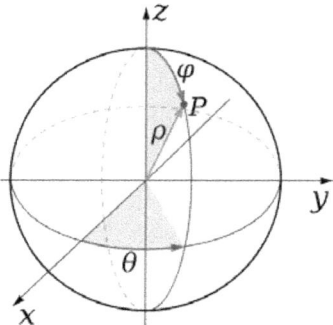

Considering a sphere of unit radius, the polar coordinate $\rho = 1, \varphi = 0°, \theta = 0°$ corresponds to the north pole. So to each coordinate corresponds a point.

Conversely, when physically considering the north pole, it can have the following polar coordinates:

$$\rho = 1, \varphi = 0°, \theta = 0°$$
$$\rho = 1, \varphi = 0°, \theta = 5°$$

$$\rho = 1, \varphi = 0°, \theta = \text{variable}$$

In practice, as θ infinite coordinates are obtained that still locate the same point.

If, on the other hand, we use an orthogonal Cartesian coordinate system in place of polar coordinates, what happens is that each point corresponds to a (x,y,z) tern, and each (x,y,z) tern corresponds to a single point.

The north pole, for example, will be uniquely determined by the single coordinate (0,0,1) such that the biunivocity condition is met.

Let us now examine locally inertial systems in a Riemann metric. In the common area, that is, where the two spaces intersect

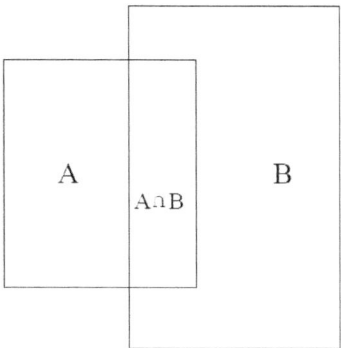

I can tie the coordinates of the portion of A with those of the intersecting portion of B through a generic function, nonlinear but invertible, for the condition set by the Riemannian metric.

$$(40c) \quad x_A = f(x_B)$$

Differentiating the nonlinear function *(40c)* gives:

$$dx_A^\mu = \frac{\partial f(x_B^\mu)}{\partial x_B^\lambda} \, dx_B^\lambda$$

$$(40d)\ dx_A^\mu = \frac{\partial x_A^\mu}{\partial x_B^\lambda}\ dx_B^\lambda$$

The same can also be expressed with other index pair

$$(40e)\ dx_A^\nu = \frac{\partial x_A^\nu}{\partial x_B^\sigma}\ dx_B^\sigma$$

Substituting *(40d)* and *(40e)* into *(40b)* gives:

$$(40f)\ d\sigma_A^2 = g_{\mu\nu}(x_A)\ \frac{\partial x_A^\mu}{\partial x_B^\lambda}\ dx_B^\lambda\ \frac{\partial x_A^\nu}{\partial x_B^\sigma}\ dx_B^\sigma$$

For space B, we can also conduct the same calculations.

Rewriting (40b) for a generic point B, however belonging to the intersection set, we have:

$$(40g)\ d\sigma_B^2 = g'_{\lambda\sigma}\ dx_B^\lambda\ dx_B^\sigma$$

Equalizing (40f) with (40g) gives:

$$g'_{\lambda\sigma}(x_B) = \frac{\partial x_A^\mu}{\partial x_B^\lambda}\frac{\partial x_A^\nu}{\partial x_B^\sigma}\ g_{\mu\nu}(x_A)$$

Substituting *x* in place of x_A and *x'* in place of x_B gives:

$$g'_{\lambda\sigma}(x') = \frac{\partial x^\mu}{\partial x'^\lambda}\frac{\partial x^\nu}{\partial x'^\sigma}\ g_{\mu\nu}(x)$$

which represents the transformation law of the metric tensor in a generic system, without the restriction of the inertial system condition, thus valid in 4-dimensional space-time with the presence of gravity.

Note how the metric tensor $g_{\mu\nu}$ no longer remains constant as the reference system varies as, on the other hand, it did in Minkowky's pseudo-Euclidean flat space, where, among other things, it took a simpler, symmetric form with 10 constant components of which 6 were null

$$g_{\mu v} = \begin{bmatrix} 1 & 0 & 0 & 0 \\ 0 & -1 & 0 & 0 \\ 0 & 0 & -1 & 0 \\ 0 & 0 & 0 & -1 \end{bmatrix}$$

Instead, it now reverts to a more generic, non-symmetrical 16-component or symmetrical 10-component form.

$$g'_{\mu v} = \begin{bmatrix} g'_{11} & g'_{12} & g'_{13} & g'_{14} \\ g'_{21} & g'_{22} & g'_{23} & g'_{24} \\ g'_{31} & g'_{32} & g'_{33} & g'_{34} \\ g'_{41} & g'_{42} & g'_{43} & g'_{44} \end{bmatrix}$$

This matrix represents the metric of curved space-time, a fundamental quantity for the development of general relativity.

If $g_{\mu v}$ is constant the space is flat (Minkowsky space, Euclidean space, etc.), conversely if $g_{\mu v}$ is not constant the space is curved.

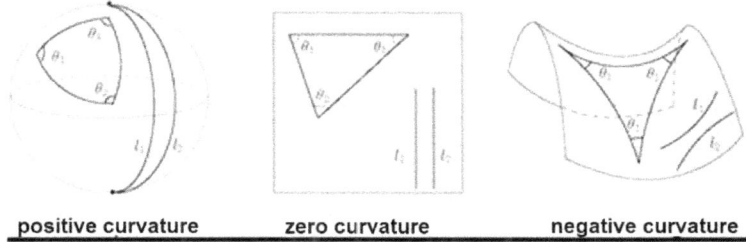

| positive curvature | zero curvature | negative curvature |

curvature in 3-dimensional space

6.2 EQUATION OF GEODESICS IN CURVED SPACE-TIME

Once we have defined the metric in space including the effects of gravitation, that is, in the field of general relativity, we can again search for the equation of geodesics in the curved space-time thus defined.

It is evident how In this case, the equation of geodesics will be different from the equation of Minkowsky M4 flat space geodesics.

Consider 2 points A and B in curved space.

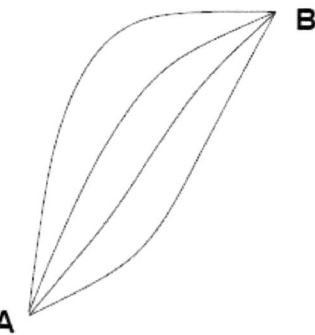

Due to what was stated in the previous paragraph, it is possible to fragment the curved spacetime into infinitesimal locally inertial spaces, where (32b) holds true

$$d\sigma^2 = g_{\mu v}(x) \; dx^\mu dx^v$$

and we can calculate the length of the geodesic (curve) through the integration of the individual locally inertial geodesics

$$s_{AB} = \int_A^B ds = \int_A^B \sqrt{g_{\mu v}(x) \; dx^\mu dx^v}$$

Assuming that the length of the geodesic is stationary with respect to infinitesimal changes of x, and denoting by δ, small changes, we have

$$\delta s_{AB} = \delta \int_A^B ds = \delta \int_A^B \sqrt{g_{\mu\nu}(x)\ dx^\mu dx^\nu} = 0$$

By parameterizing the x coordinates in s, x(s)

$$\delta s_{AB} = \delta \int_A^B \sqrt{g_{\mu\nu}(x)\ dx^\mu dx^\nu}\, \frac{ds}{ds} =$$

$$= \delta \int_A^B \sqrt{g_{\mu\nu}(x)\ \frac{dx^\mu}{ds}\frac{dx^\nu}{ds}}\, ds = 0$$

By performing laborious calculations omitted here, we obtain the equation of geodesics in curved space, suitable for describing the motion of gravities in general relativity:

$$(40h) \quad \frac{d^2 x^\lambda}{ds^2} + \Gamma_{\mu\nu}^\lambda \frac{dx^\mu}{ds}\frac{dx^\nu}{ds} = 0$$

Where the operator $\Gamma_{\mu\nu}^\lambda$ is the Christoffel symbol equal to

$$\Gamma_{\mu\nu}^\lambda = \frac{1}{2} g^{\lambda\sigma}\left(\frac{\partial g_{\sigma\mu}}{\partial x^\nu} + \frac{\partial g_{\sigma\nu}}{\partial x^\mu} - \frac{\partial g_{\mu\nu}}{\partial x^\sigma}\right)$$

(40h) in pseudoeuclidean or Minkowsky space (M4), in view of the condition of inertial systems, where the prime derivatives of the metric tensor $g_{\mu\nu}$ are zero, the Christoffel symbol is zero $\left(\Gamma_{\mu\nu}^\lambda = 0\right)$ and (40h) reduces to

$$(40i) \quad \frac{d^2 x^\lambda}{ds^2} = 0$$

The latter represents the equation of a line in 4-dimensional space.

In Euclidean space, however, *(40i)* represents, more simply, the most classical straight line in 3-dimensional space.

This result allows us to place a further consideration on *(40h)*: the first term represents the inertial rate while the second term represents the gravitational rate, between them quite distinct.

Ultimately, the motion of a grave in space with gravity always follows a geodesic in space-time, which is equivalent to a line in Minkowksy space-time, suitably distorted by gravity; all quantitatively represented by the two terms in equation *(40h)*.

The action of the gravitational field thus manifests itself as a distortion of Minkowski space-time.

For the Earth, given the small value of mass, this distortion is practically imperceptible.

The Earth in its motion, follows a geodesic and, as is well known, moves along an ellipse.

Based on observation, one might think that the Sun causes such a strong distortion with its gravity that it changes the straight trajectory into an ellipse.

However, we made a mistake, neglecting the fourth dimension time, in visualizing geodesics.

So let's understand it better by also considering the fourth dimension and representing the earth's trajectory in a space-time reference system instead of just space.

With the addition of the time dimension, the Earth's trajectory turns out to have a geometric shape of a kind of helix with a very elongated elliptical projection and therefore very close to a straight line.

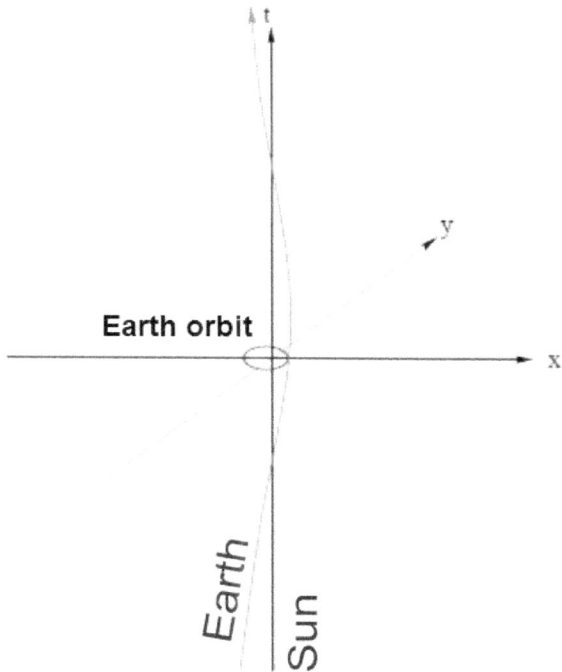

It is indeed a geodesic of space-time slightly deformed by a straight line due to solar gravity. The ellipse that describes its orbit is nothing more than the projection of this geodesic, in 3-dimensional space.

It is pointed out that when we say that a body in a gravitational field moves along geodesics, we always mean geodesics in a 4-dimensional space-time.

6.3 TENSORS AND SPACE-TIME

The relationship between gravitational field and curvature will necessarily have to be a mathematical entity that correlates representative parameters of curvature (Riemann tensor) with other parameters involving masses (energy-impulse tensor).

In order to formulate the field equations, it will be necessary to use new entities, called tensors, that will allow us to describe the ways in which physical or geometric entities vary with the change of coordinates, that is, that will allow us to mathematically represent the expected deformation of space-time.

A tensor in one local reference system is related by appropriate linear function to the corresponding tensor of another local reference system, which is then easily determined.

Let us briefly understand what a tensor is.

Tensors, are an algebraic evolution of vectors.

A vector or countervector is an element of a vector space, consisting of an n-uple of values, i.e., a 1xn matrix (1 row x n columns)

In an n-dimensional space we have:

$$\mathbf{V}= (V^1 , V^2 , V^3 , V^4 , \ldots\ldots V)^n$$

More simply with index notation we can denote the components of a vector as

V^α with α = 1,2,3,4, ... n (n.b. index placed at the top)

The components of a "vector" are called countervariant components.

Given a vector, the corresponding "co-vector" is derived from a linear scalar function of the vector.

The co-vector can also be understood as a mathematical operator, which for each vector gives rise to a number.

A co-vector unlike a vector is represented with the index at the bottom

$$\boldsymbol{P}_\alpha = (P\,,P\,,P_{123}\,,P_4\,,\,.....P\,)_n$$

Or

$\boldsymbol{P}_\alpha$ with α = 1,2,3,4, ... n (n.b. index placed at the bottom)

The components of a covector are referred to as covariant components.

The tensor is nothing more than an appropriate linear scalar function that binds vectors and co-vectors.

More simply, we can define Tensor T as an operator that takes h vectors and k covectors as input, and transforms them into a scalar.

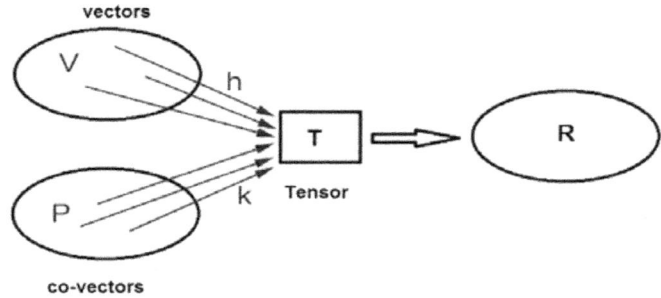

The symbology for indicating tensors is derived immediately from their constituents

$$\mathbf{T}(\mathbf{V}^h,\mathbf{P})_k = T^h_k$$

Where h represents the number of input vectors and k the number of covectors.

The rank or order of the tensor is given by the sum of the input vectors and covectors, and is denoted by

$$\binom{h}{k} \quad oppure \quad r = h + k$$

The rank is independent of the number of components of the vector or covector, but depends solely on the number of input vectors and covectors.

A Tensor of rank 0, $\binom{0}{0}$, has zero input vectors and zero input covectors, so it is a scalar.

A Tensor of rank 1, $\binom{0}{1}$ o $\binom{1}{0}$, has an input vector, or in the second case an input covector, so it is a vector or covector consisting of a n-uple.

A Tensor of rank 2, $\binom{0}{2}$ o $\binom{2}{0}$ o $\binom{1}{1}$, has zero input vectors and two covectors in the first case, two input covectors and zero vectors in the second, and finally one vector and one covector in the last case, remaining in all three cases represented by a square matrix nxn, with n equal to the number of components of the vector or covector.

A tensor of rank 3, on the other hand, is a cubic lattice nxnxn.

In all cases n represents the number of components of the input vectors/covectors.

Unfortunately, only tensors of rank 0, 1 and 2 can be treated with the usual rules of matrix calculus, since they are represented as numbers, vectors and matrices.

As an example, let us analyze a vector $V(x^1, x^2, x^3, x^4)$ in Minkowsky space M4 together with another vector $V'(x', x'^{12}, x'^3, x'^4)$.

In this case we have that the tensor **T** of the vectors **V** and **V'** will be of rank 2, due to the absence of covectors and having only 2 input vectors $\binom{2}{0}$. The resulting tensor will be a 4x4 matrix

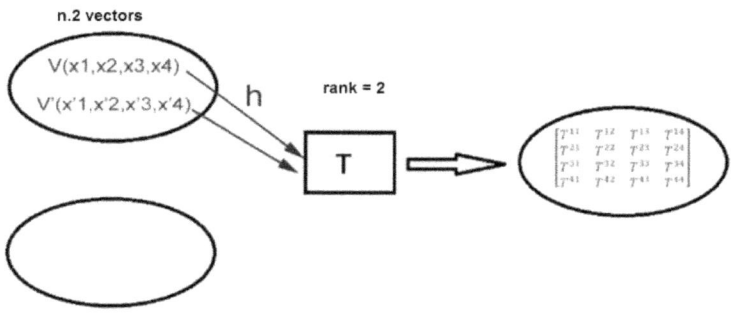

$$\mathbf{T(V,V')} = [T^\alpha T^\beta] = \begin{bmatrix} T^{11} & T^{12} & T^{13} & T^{14} \\ T^{21} & T^{22} & T^{23} & T^{24} \\ T^{31} & T^{32} & T^{33} & T^{34} \\ T^{41} & T^{42} & T^{43} & T^{44} \end{bmatrix}$$

Note that the tensor components have the indices at the top (superscript) as they refer to vectors.

Conversely, if we had covectors as input, the corresponding tensor should have had the components represented with the indices placed at the bottom (subscript).

Ultimately, the introduction of tensors, in General Relativity, is necessary in order to be able to handle the change of reference systems, because of the property of binding according to a linear and homogeneous function between the different local reference systems. Tensors transform according to the countervariant and covariant schemes following any coordinate transformation.

For the formulation of general relativity, the Riemann tensor (R) is used as the geometric entity containing precisely all information regarding the curvature of space-time.

If the Riemann Tensor is zero at a point, then the space is flat at that point.

The Riemann tensor is of the type $R^{\alpha}_{\beta\mu\sigma}$, born from an input vector and three covectors, so it is a tensor of rank 4 of type $\binom{1}{3}$, which as we know does not enjoy the properties of matrix calculus.

Beginning with that tensor and taking advantage of properties on tensor calculus, the tensor can be contracted (contraction or trace rule) by performing sums over equal or different pairs of indices, such that the resulting contracted tensor becomes two orders less, that is, of rank equal to *r=h+k-2*.

By introducing the fully covariant Riemann tensor $R_{\alpha\beta\mu\sigma} = g_{\alpha\mu}R^{\alpha}_{\beta\mu\sigma}$ (indices all down) and contracting the same with the metric tensor $g^{\alpha\mu}$ at $\boldsymbol{\alpha} = \boldsymbol{\mu}$, we give rise to the Ricci tensor in the form $R_{\beta\sigma} = R_{\sigma\beta}$ (symmetric tensor $R_{\beta\sigma} = g^{\alpha\mu}R_{\alpha\beta\mu\sigma}$) of rank 2 (4x4 symmetric matrix in 4-dimensional space).

By further contracting Ricci's tensor $R_{\beta\sigma}$ on $\beta = \sigma$ we obtain the Ricci scalar (R = $g^{\beta\sigma}R_{\beta\sigma}$).

It is understood that the statement on tensor calculus, is of the summarizing type, while those who wish to go deeper will have to turn to books that deal rigorously with tensor exposition (in Bibliography: Tensors made easy by G. Bernacchi and Relativity by V.Barone).

We summarize the steps followed so far, preparatory to the subsequent formulation of the field equations:

1) Space-time is fragmented into many small locally inertial (or locally flat) spaces where the theory of special relativity remains valid, that is, where "locally" the influence of gravity is negligible;

2) the metric tensor $g_{\mu\nu}$ is no longer constant in space-time as it was in Monkowsky's M4 space, presenting variable components that are functionally related to the coordinates of the reference systems considered;

3) The equation of geodesics in the presence of gravity, has two terms: the first related to inertial geodesics and the second related to gravitational distortion.

The action of the gravitational field thus manifests itself as a distortion of Minkowski space-time.

4) The use of tensors and related tensor calculations is introduced because of the property of binding according to a linear and homogeneous function between different local reference systems.

Unlike Newton's gravitation, the gravitational field in general relativity will no longer be described by a scalar field (the gravitational potential), but by a 10-component tensor field (by symmetry), as it is related to the metric tensor $g_{\mu\nu}$ which has precisely 10 components.

Thus Newton's gravitational potential is identified with the metric tensor of the relativistic case.

The prime derivatives of the metric tensor, in analogy to the classical case, represent the acceleration of gravity.

The above prime derivatives are identified by the Christoffel symbol $\Gamma^{\lambda}_{\mu\nu}$, and are 40-component objects, as opposed to the only 3 components in (x,y,z) of the acceleration of gravity (g) in the classical case.

The second derivatives of the metric tensor, on the other hand, represent the curvature of space-time and have to do with the tidal forces, which are those forces induced by the change in the acceleration of gravity at different neighboring points in space due to the change in distance from the distortion-generating massive body.

6.4 EINSTEIN'S FIELD EQUATIONS

The universe extends in a 4-dimensional warped space-time, where all matter moves according to geodesics of space-time as a function of matter density, energy and pressure, all succinctly represented by the "**Einstein field equations,**" an expression of the conclusive result of the theory of general relativity, distinguished by the simplicity of matter/energy and curvature coupling:

$$(41) \quad R_{\mu\nu} - \frac{1}{2}g_{\mu\nu}R + \Lambda g_{\mu\nu} = \frac{8\pi G}{c^4}T_{\mu\nu}$$

The values placed to the left of the equation represent the properties of "warped space-time," while the quantities placed to the right represent "matter-energy density."

The equation can be read as:

The deformation of space-time is proportional to the amount of matter-energy.

Space-time tells matter how to move, and matter tells space-time how to curve (John Archibald Wheeler).

Let us analyze the components of the equation:

$$R_{\mu\nu}$$ = Ricci curvature tensor

A Tensor is from a physical point of view, an object (sequence of numbers, matrix) defined from a vector space even with multiple dimensions that is not dependent on a particular reference system.

Ricci's tensor measures how the volume form of the variety differs locally from the usual Euclidean volume form, that is, it measures its curvature.

The hedgehog tensor is covariant of order or rank 2 (two indices at the bottom of the R).

$g_{\mu\nu}$ = Metric tensor

The metric tensor describes the metric of space-time and is a symmetric tensor of rank 2, in the matrix form 4x4 (3 spatial dimensions + 1 temporal dimension), and is apt to define the notions of distance, angle, length of a curve, geodesics, curvature.

Said tensor for example in Euclidean space R³ (n=3) reduces to an identity matrix 3x3

$$g_{\mu\nu} = \begin{bmatrix} 1 & 0 & 0 \\ 0 & 1 & 0 \\ 0 & 0 & 1 \end{bmatrix}$$

Which precisely is representative of the invariance of the distance entity referred to in relation (30)

$$ds^2 = (dx^2 + dy^2 + dz^2)$$

written in Einstein's notation

$$ds^2 = g_{\mu\nu} \, dx^\mu dx^\nu$$

which we recall is equivalent to

$$ds^2 = \sum_{\mu,\nu=1}^{3} \left(g_{\mu\nu} \, dx^\mu dx^\nu \right)$$

In the Minkowski space of type R⁴ the metric tensor, on the other hand, is as follows:

$$g_{\mu\nu} = \begin{bmatrix} 1 & 0 & 0 & 0 \\ 0 & -1 & 0 & 0 \\ 0 & 0 & -1 & 0 \\ 0 & 0 & 0 & -1 \end{bmatrix}$$

Which component of the famous equation (31) invariant constituent in such a space

$$d\sigma^2 = c^2 dt^2 - (dx^2 + dy^2 + dz^2)$$

written in Einstein's notation,

$$ds^2 = g_{\mu\nu} \, dx^\mu dx^\nu$$

which, remembering again the sum rule for equal indices, is equivalent to

$$ds^2 = \sum_{\mu,\nu=1}^{4}\left(g_{\mu\nu} \, dx^\mu dx^\nu\right)$$

where the components

$$dx^1 = c \, dt$$
$$dx^2 = dx$$
$$dx^3 = dy$$
$$dx^4 = dz$$

R = Scalar curvature

Scalar curvature is defined from the Ricci curvature tensor, a trace of which is provided through the use of the metric tensor **g**. In analytical terms

$$R = g^{\mu\nu} \, R_{\mu\nu}$$

Scalar curvature is not a matrix but a scalar function.

Λ = Cosmological constant

Being placed on the left side of the equation, the cosmological constant was initially understood as a property of space-time.

This value was introduced by Einstein to balance the two members of the equation and make the universe static, which was otherwise found to be of a dynamic type (expansion or contraction).

Following the observations of Hubble in 1929 it was shown that the universe is expanding, and Einstein himself felt that the introduction of said constant was "his biggest mistake."

Instead, it seems he had been right, in fact today the cosmological constant has been reevaluated to explain the acceleration of the expansion of the universe, with a role of large-scale anti-gravitational force, represented by dark energy, such as vacuum energy and quintessence.

$T_{\mu\nu}$ = Stress–energy tensor

Also referred to as Impulse Energy Tensor, it describes the flow of momentum (quadrimpulse) and energy through a hyper-surface .

It is covariant of order or rank 2 (two indices down the T).

G , c = Newton's gravitational constant and the speed of light in vacuum

The first is the well-known and already well-known gravitational constant introduced by Newton, and the second is the constant speed of light in a vacuum.

6.5 QUALITATIVE ANALYSIS OF FIELD EQUATIONS

Einstein's Equations from a qualitative point of view show that the curvature of space-time is proportional to mass density.

A massive body, with high mass density, causes a relevant space-time distortion.

For the analogy between mass and energy, introducing the mass density ρ, which turns out to be proportional to the tensor **T**,

$$\rho \sim \frac{T_{00}}{c^2}$$

Considering that the first term of (41) represents the curvature of space-time,

$$curvature\ space{-}time \sim \frac{G\rho}{c^2}$$

and that the ratio

$$\frac{G}{c^2} = 7.4 \cdot 10^{-30} \text{m}\, Kg^{-1}$$

we obtain that the space-time curvature is proportional to the mass density minus a very small factor of the order of 10^{-30},

$$space - time curvature \sim 7.4 \cdot 10^{-30}\, \rho$$

This result leads to the expected conclusion that in order to have appreciable values of space-time deformation, it is necessary to have very high mass densities.

The field equation (41), provides different solutions based on the boundary conditions considered and the specific cases addressed.

In particular, it makes it possible to predict important phenomena, analyzed in the following paragraphs, such as:

- the gravitational deflection of light

- The precession of the perihelion of planetary orbits
- the Gravitational redshift (redshift)
- gravitational waves

6.6 GRAVITATIONAL DEFLECTION OF LIGHT

During a total eclipse, you can see some of the brightest stars in the sky, which normally the sunlight prevents you from seeing.

By comparing a photographic plate taken through the telescope during the eclipse with one of the same region of the sky taken at night, a difference in the position of the stars can be seen.

The effect is greater the closer the stars are to the direction of the Sun.

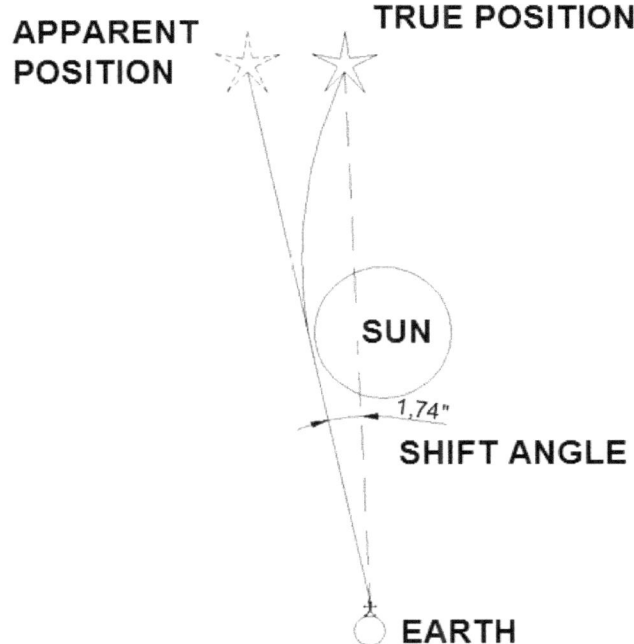

The effect is due to the gravitational deformation of space by the massive sun.

The deviation of light from the stars closest to the Sun's edge results in about 1.75". This value can be derived both experimentally and by applying Einstein's theory of Relativity, through solving the field equations.

Analytically solving (41) for given boundary conditions, we obtain

$$\alpha = \frac{4\,G\,M}{R\,c^2}$$

Substituting the corresponding values of G, c, R and M, Gravitational constant, speed of light in vacuum, radius and solar mass, respectively, yields the expected value of

$$\alpha = 8.47 \cdot 10^{-6}\ rad = 1.75"$$

The first experimental verification was performed during the May 29, 1919 eclipse by Arthur Eddington, simultaneously in the cities of Sobral (Brazil) and in São Tomé and Príncipe on the west coast of Africa, during a total eclipse.

As a result of this verification, Einstein was able to have proof of the truth of his theoretical assertions about the deflection of light. When asked what he had done in case of a negative finding of the Theory of Relativity, Einstein replied with the famous line, "I would therefore feel sorry for the dear man, but the theory is correct anyway."

6.7 PERIHELION PRECESSION OF PLANETARY ORBITS

The solution of (41) allows us to describe the motion of the planets around the Sun, which as is well known their orbit is an ellipse.

The planet MERCURY, however, while maintaining its elliptical orbit, undergoes perihelion precession, which cannot be predicted by Newtonian gravity alone.

From calculation with Newtonian gravitational theory alone, the value obtained was reduced from what was expected from experimental measurements.

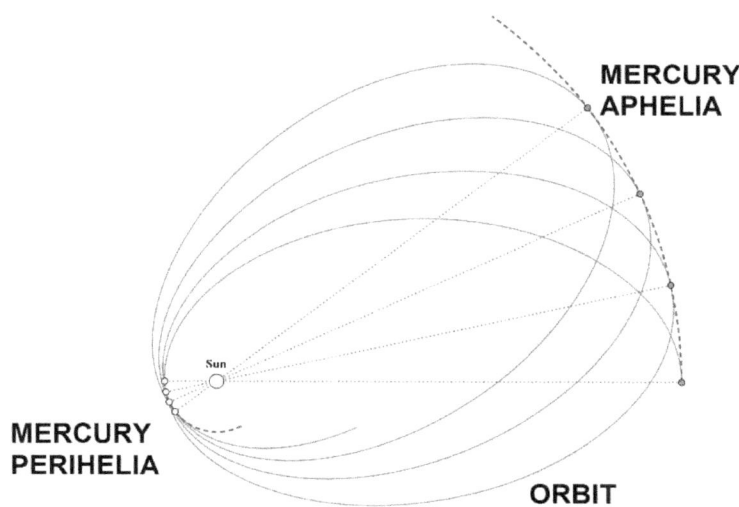

Said difference is easily recovered by comparing the solution of (41), considering only the solar mass and the planet in question, with Newton's equations. A value of 0.1036" (in arcseconds) is derived.

The difference recovered in terms of secular precession, given that the planet Mercury has a revolution period of 87.8 days, has a value of:

$$\delta_{100} = \frac{\delta \cdot 100 \cdot 365.25}{87.8} = 43''$$

The value obtained represents, the difference of what actually is the secular advance of perihelion (574") and what is calculated by Newtonian theory (532"), taking into account the perturbations of other planets.

This result is in perfect agreement with the experimentally measured values.

6.8 REDSHIFT

The term REDSHIFT or "RED SHIFT" refers to the phenomenon whereby the light emitted by an object has a longer wavelength than it had at emission.

It is the analogue of the DOPPLER effect for sound waves.

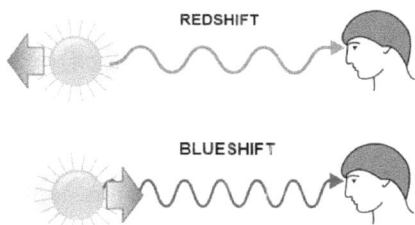

By analyzing the electromagnetic spectrum, it can be seen that the classification of an electromagnetic wave starting from the visible, as the wavelength increases, shifts to the infrared (toward the left), from which the respective name descends.

Otherwise as the wavelength decreases, the classification shifts from the visible to the ultraviolet (toward the right).

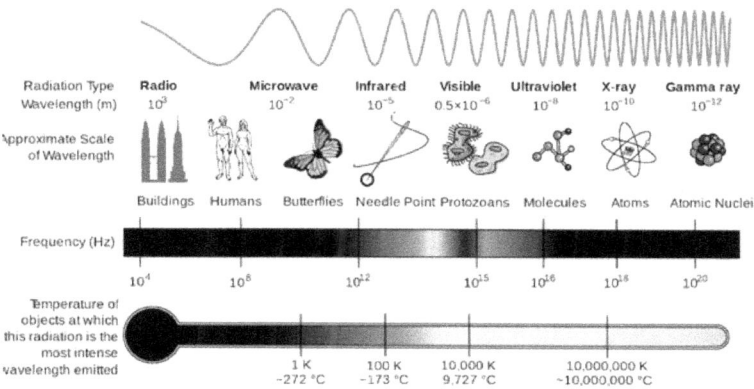

Knowing the wavelength emitted by a celestial body, by measuring the wavelength that reaches us, we can know whether it is moving away from us or approaching us.

It should be noted that it is not the wavelength of the photon and its energy ($E=h\,\nu$) that varies, but rather the arrival time that lets a change in wavelength appear.

So we talk about apparent wavelength variation.

This effect is most useful in determining the receding velocities of celestial bodies measured relatively from Earth.

Observing the universe, we come to the conclusion that it is expanding in every part of it, with increasing speed as the radius of hyperspace increases, due to the increase in the acceleration of expansion as distance increases.

Increasing the speed of expansion will result in a completely dark universe; in fact, when the speed of expansion is close to that of light, light will escape the perception of the universe's hosts.

The measurement of REDSHIFT is not easy to account for, as the value obtained must be properly corrected with the value of gravitational REDSHIFT, which is quite different from the former.

According to general relativity in the vicinity of a massive body there is time dilation. Time dilation leads to a larger period value of the T wave, resulting in a smaller frequency value and still a longer wavelength.

Gravitational REDSHIFT depends on the mass of the generating body.

Through appropriate solutions of Einstein's field equations it is possible, to correlate the Redshift value with the mass of the inspected body.

The relationship, simplified, that links the two values is as follows:

$$1 + z = \frac{1}{\sqrt{1 - \dfrac{2\,G\,M}{r\,c^2}}}$$

That for small values of z and performing arrested Taylor series development to first order, we obtain:

$$z = \frac{G\,M}{r\,c^2}$$

Where

$z = n - \ \nu' =$ *difference in frequencies*

G = *gravitational constant*

c = *speed of light*

r = *radial distance*

Recently, as a result of a study coordinated by the University of California at Los Angeles (UCLA), led by researchers Andrea Ghez and Tuan Do, published in Science, a gravitational redshift was observed in the light emitted by the star S0-2, which is the closest star to the center of our galaxy, orbiting the supermassive black hole Sagittarius A*, whose mass is four million times that of the Sun and is located at the center of the Milky Way.

The star S0-2, takes 16 years to make an orbit around Sagittarius A*, which is why the scientists have carried out more than 20 years of observations at the Keck Observatory (the Keck telescopes) in Hawaii through a spectrograph built at UCLA and used the most recent 2018 data for analysis.

A red shift was observed when the star was in close proximity to the black hole during its 16-year orbit. The team made repeated observations during crucial periods to detect the phenomenon, analyzing photons (the particles of light) as they traveled from the star to Earth. The wavelengths examined were found to depend not only on the star's speed but also on the amount of energy it took to escape the supermassive black hole's extremely powerful gravitational field.

The result of the study is, once again, a confirmation of the predictions of Einstein's theory of general relativity.

6.9 GRAVITATIONAL WAVES

The field equations also predict solutions of the wave type, of the type

$$h_{\mu v} = e_{\mu v}(k)e^{-ik\,x}$$

However, the power of gravitational waves generated by accelerated masses is small and therefore difficult to capture or directly observe.

Gravitational waves generated by the planet Earth in the course of its motion of revolution have a power of about 200 watts, clearly difficult to detect.

In the same way that a wave in a fluid ripples the surface, propagating and causing the front to oscillate around equilibrium values, so gravitational waves, propagating at the speed of light, change the space-time structure of nearby points with oscillation around reference values.

Gravitational Waves were predicted in the early 1900s by Einstein in the Theory of General Relativity.

In the 1970s, deduced by U.S. scientists Russell Hulse and Joseph Taylor by observing a pair of pulsars orbiting each other. In the course of the observation, it emerged that these compact stars slowed down as they approached each other, precisely because of the loss of energy in the form of gravitational waves. The first certain observation of gravitational waves, however, did not occur experimentally until September 2015.

BIBLIOGRAPHY

Dialogo sopra i due massimi sistemi del mondo, Galileo Galilei, A. Beltrán Marí (a cura di), BUR Biblioteca Univ. Rizzoli (1 dicembre 2014)
Einstein. «Sottile è il Signore...». La scienza e la vita di Albert Einstein, Abraham Pais, T. Cannillo (a cura di), G. Belloni (Traduttore), Bollati Boringhieri (2 febbraio 2012)
Le due relatività. Gli articoli originali del 1905 e 1916, Albert Einstein, E. Sagittario (Traduttore), A. Pratelli (Traduttore), Bollati Boringhieri (5 novembre 2015)
La relatività, Alberto Bandini Buti, Sandit Libri (5 marzo 2007)
Manuale di relatività ristretta. Per la Laurea Triennale in fisica, Gasperini, Maurizio, Springer Verlag; 2010 ed. (21 aprile 2010)
Relatività generale e teoria della gravitazione, Maurizio Gasperini, Springer Verlag; 2 edizione (10 settembre 2014)
La Teoria Della Relativita' Speciale e Generale: Esposizione Semplificata, Conrad, C. G., Createspace Independent Pub; 1 edizione (25 febbraio 2015)
Relatività. Guida illustrata molto speciale. Ediz. Illustrata, Sander Bais, A. Migliori (Traduttore), Dedalo (30 aprile 2008)
Relatività. Principi e applicazioni, Vincenzo Barone, Bollati Boringhieri (2 dicembre 2004)
Tensori fatti facili con Problemi, Giancarlo Bernacchi, lulu.com (9 dicembre 2016)
Quotes on the cover:
https://www.brainyquote.com/quotes/saint_augustine_108119
https://www.scientificamerican.com/article/pioneering-physicist-john-wheeler-dies/